MEDICAL
INTELLIGENCE
UNIT

Hepatitis Delta Virus

Hiroshi Handa, M.D., Ph.D.
Graduate School of Bioscience and Biotechnology
Tokyo Institute of Technology
Nagatsuta, Yokohama, Japan

Yuki Yamaguchi, Ph.D.
Graduate School of Bioscience and Biotechnology
Tokyo Institute of Technology
Nagatsuta, Yokohama, Japan

LANDES BIOSCIENCE / EUREKAH.COM
GEORGETOWN, TEXAS
U.S.A.

SPRINGER SCIENCE+BUSINESS MEDIA
NEW YORK, NEW YORK
U.S.A.

HEPATITIS DELTA VIRUS

Medical Intelligence Unit

Landes Bioscience / Eurekah.com
Springer Science+Business Media, Inc.

ISBN: 0-387-32230-2   Printed on acid-free paper.

Springer Science+Business Media, Inc., 233 Spring Street, New York, New York 10013, U.S.A.
http://www.springer.com

Please address all inquiries to the Publishers:
Landes Bioscience / Eurekah.com, 810 South Church Street, Georgetown, Texas 78626, U.S.A.
Phone: 512/ 863 7762; FAX: 512/ 863 0081
http://www.eurekah.com
http://www.landesbioscience.com

Printed in the United States of America.

9 8 7 6 5 4 3 2 1

Library of Congress Cataloging-in-Publication Data

Hepatitis delta virus / [edited by] Hiroshi Handa, Yuki Yamaguchi.
     p. ; cm. -- (Medical intelligence unit)
  Includes bibliographical references and index.
   ISBN 0-387-32230-2 (alk. paper)
   1. Delta-associated agent. 2. Delta infection. I. Handa, H. (Hiroshi), 1946- II. Yamaguchi, Yuki. III. Series: Medical intelligence unit (Unnumbered : 2003)
   [DNLM: 1. Hepatitis Delta Virus--genetics. 2. Hepatitis D. 3. Hepatitis Delta Virus--physiology. 4. Virus Replication--physiology. QW 170 H5298 2006]
   QR201.H46H453 2006
   616.3'623019--dc22
                         2006005520

CONTENTS

EDITORS

Hiroshi Handa
Graduate School of Bioscience and Biotechnology
Tokyo Institute of Technology
Nagatsuta, Yokohama, Japan
Chapter 6

Yuki Yamaguchi
Graduate School of Bioscience and Biotechnology
Tokyo Institute of Technology
Nagatsuta, Yokohama, Japan
Chapter 6

CONTRIBUTORS

John L. Casey
Department of Microbiology
  and Immunology
Georgetown University Medical Center
Washington, District of Columbia,
  U.S.A.
Email: caseyj@georgetown.edu
Chapter 5

Nobuyuki Enomoto
First Department of Internal Medicine
University of Yamanashi
Yamanashi, Japan
Chapter 1

Stephanos J. Hadziyannis
Department of Medicine
  and Hepatology
Henry Dunant Hospital
Athens, Greece
Email: hadziyannis@ath.forthnet.gr
Chapter 7

Michael M.C. Lai
Department of Molecular Microbiology
  and Immunology
University of Southern California
Keck School of Medicine
Los Angeles, California, U.S.A.
and
Institute of Molecular Biology
Academia Sinica
Taipei, Taiwan
Email: michlai@usc.edu
Chapter 4

Kazuyoshi Nagayama
Department of Gastroenterology
  and Hepatology
Tokyo Medical and Dental University
Tokyo, Japan
Chapter 1

Camille Sureau
CNRS
Laboratoire de Virologie Moléculaire
INSERM U76
Institut National de la Transfusion
  Sanguine
Paris, France
Email: csureau@ints.fr
Chapter 2

John M. Taylor
Fox Chase Cancer Center
Philadelphia, Pennsylvania, U.S.A.
Email: jm_taylor@fccc.edu
Chapter 3

Dimitrios Vassilopoulos
Athens University School of Medicine
Hippokration General Hospital
Academic Department of Medicine
Athens, Greece
Email: dvassilop@med.uoa.gr
Chapter 7

Hideki Watanabe
Department of Gastroenterology
   and Hepatology
Tokyo Medical and Dental University
Tokyo, Japan
Chapter 1

Mamoru Watanabe
Department of Gastroenterology
   and Hepatology
Tokyo Medical and Dental University
Tokyo, Japan
Chapter 1

Jaw-Ching Wu
Division of Gastroenterology
Taipei Veterans General Hospital
Institute of Clinical Medicine
National Yang-Ming University
Taipei, Taiwan
Email: jcwu@vghtpe.gov.tw
Chapter 8

Tsuyoshi Yamashiro
First Department of Internal Medicine
University of Ryukyus
Okinawa, Japan
Chapter 1

PREFACE

Since its discovery in 1979, HDV has occupied a unique position in virus taxonomy. It does not belong to any of the established viral family but constitutes its own genus, deltavirus, whereas it does have significant similarity to viroids, subviral agents of higher plants. HDV RNA genome is smaller than any known animal virus genome, so small that it encodes only a single protein. Therefore, its propagation is largely dependent on factors supplied by host and another virus, hepatitis B virus (HBV). For example, HDV makes use of HBV's surface antigens for envelope proteins. HDV replicates through RNA-dependent RNA synthesis by cellular DNA-dependent RNA polymerase(s). RNA editing by cellular enzyme(s) and RNA cleavage by viral ribozymes are also involved in the viral life cycle. From a medical point of view, patients infected with both HBV and HDV tend to develop more severe clinical symptoms than those infected with HBV alone. All these features make HDV unique and attractive, and its research over the last two decades has resulted in a number of findings that have wide implications beyond the immediate subject.

This book concisely describes various aspects of HDV, from basics to cutting-edge research, from medicine to molecular virology and biology. Chapters were written by internationally renowned scientists. We want to take this opportunity to thank all the authors who generously contributed. We hope their conscientious efforts will have made this book useful to broad readers for many years to come. We would also like to acknowledge the expert assistance of Cynthia Conomos and Sara Lord at Landes Bioscience.

Hiroshi Handa, M.D., Ph.D.
Yuki Yamaguchi, Ph.D.

CHAPTER 1

Genotype of Hepatitis Delta Virus

Nobuyuki Enomoto,* Hideki Watanabe, Kazuyoshi Nagayama,
Tsuyoshi Yamashiro and Mamoru Watanabe

Classification of HDV Genotype

Hepatitis delta virus (HDV) is a defective virus that requires hepatitis B virus (HBV) surface antigen for virion assembly and infection,[1] and contains a negative single stranded circular RNA genome of 1.7 kilobases.[2,3] HDV is classified into three genotypes (genotype I, II and III) based on genetic sequence analysis (Fig. 1).[4] Genotype II shows approximately 75% homology to genotype I, and genotype III shares about 60 to 65% homology with genotype I and II. There are many variants within each genotype. Especially, HDV genotype II is further divided into two types (i.e., IIa and IIb), with 77% nucleotide homology between the complete sequences of genotype IIa and IIb.[5] The nucleotide homology between genotype IIb and IIb-M, the newly identified IIb variant, is 88-90%.[6] Similarly, IIa variant was recently found in Siberia (IIa-Yakutia), which in comparison with IIa shows a similar degree of genetic differences.[7] These genotypes show different geographical distributions and clinical pictures, which is thought to be caused by functional differences of genotype-specific sequences of HDV-RNA as well as HDAg protein.

Geographical Distribution of HDV Genotype

Genotype I has been identified in most areas of the world and represented by many different isolates (Fig. 1).[8] Genotype II is confined to East Asia (mainly Siberia, Japan, and Taiwan),[9] in contrast to the ubiquitous global distribution of genotype I. Genotype IIb was first identified in Taiwan,[5] and was subsequently reported among patients from the Miyako Islands,[10] one of the nearest Japanese islands to Taiwan. Recently, a new genetic variant of HDV genotype IIb (IIb-M) was identified.[6] Genotype III is isolated to the northern part of South America, and is closely associated with fulminant hepatitis.[4]

Clinical Significance of HDV Genotype

HDV genotypes are known to affect the pathogenesis and diverse clinical pictures of HDV infection.[4,7,9] Genotype I causes hepatic diseases ranging from mild to severe, often with the aggressive hepatitis and frequently associated with liver cirrhosis (LC) and hepatocellular carcinoma (HCC). On the other hand, genotype II is generally associated with a more favorable outcome than genotype I.[9] A IIa variant recently reported in Yakutia, Siberia, Russia also causes

*Corresponding Author: Nobuyuki Enomoto—First Department of Internal Medicine,
University of Yamanashi, Shimokato, Tamaho, Yamanashi 409-3898, Japan.
Email: enomoto@yamanashi.ac.jp

Hepatitis Delta Virus, edited by Hiroshi Handa and Yuki Yamaguchi.
©2006 Landes Bioscience and Springer Science+Business Media.

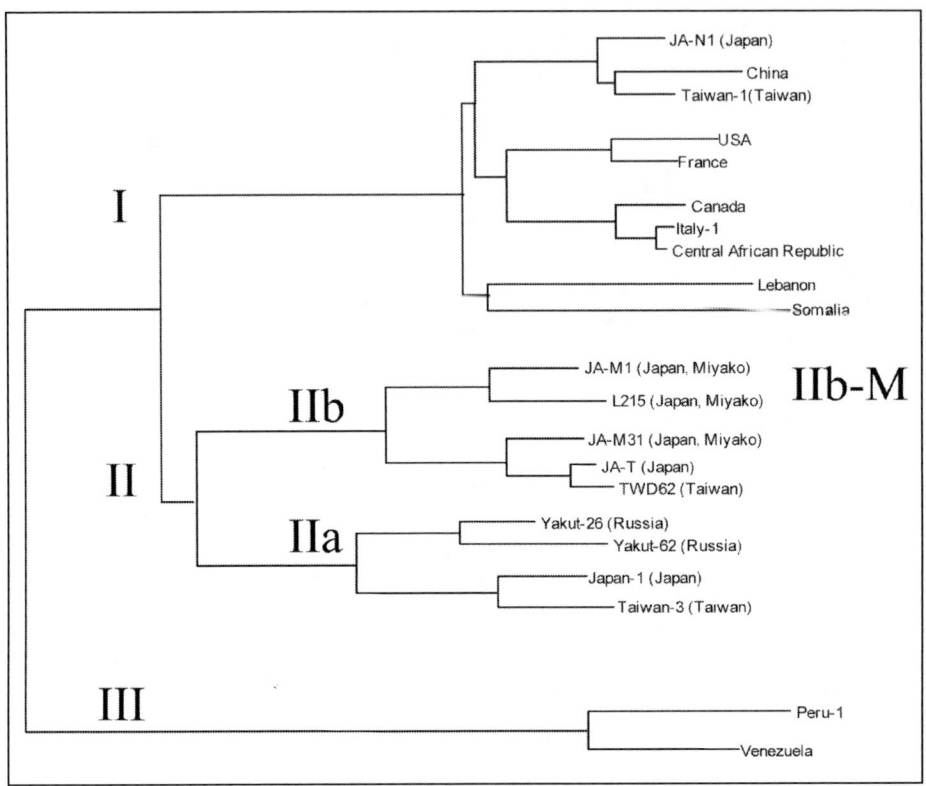

Figure 1. Phylogenetic tree analysis of HDV isolates. Sources of isolates are as follows: TWD62 (AF018077), Taiwan-3 (U19598), Taiwan-1 (M92448), Yakut-26 (AJ309879), Yakut-62 (AJ309880), Japan-1 (X60193), Lebanon (M84917), Somalia (U81988), China (X77627), USA (M28267), France (D01075), Italy-1 (X04451), Canada (AF098261), Central African Republic (AJ000558), Peru-1 (L22063), Venezuela (AB037948), JA-M1 (AF309420), JA-M31(AB118841), JA-T (AB118847) were sequenced in this study. (GenBank accession number).

a severe hepatitis comparable to genotype I in this cohort.[7] Genotype III is associated with fulminant hepatitis.[4] These findings strongly suggest that the genetic structure of HDV can profoundly influence the pathogenesis of liver injury in HDV infection.

In Japan, chronic HDV infection is endemic in the Miyako Islands where genotype IIb and IIb-M are found, and their clinical pictures differed despite relatively uniform clinical backgrounds including virological factors of HBV.[10,11] Most of the patients with chronic HDV genotype IIb infection were asymptomatic carrier (ASC) or chronic hepatitis (CH) and none were at the liver cirrhosis (LC) or hepatocellular carcinoma (HCC) stage. In contrast, about half of patients with genotype IIb-M were in the CH and LC stages, respectively, and none of them were ASC.[6] These findings indicate that patients with genotype IIb-M are more likely to progress to LC and HCC than those with genotype IIb, and that differences in HDV genotype could cause the different clinical pictures observed in this population.

In general, the genetic structure responsible for clinical features could not be readily determined because the genetic differences between the different genotypes are too diverse as seen in Figure 2. In contrast, despite the different clinical pictures between IIb and IIb-M, the genetic differences are small enough to enable the definition of the genetic features of HDV pathogenesis

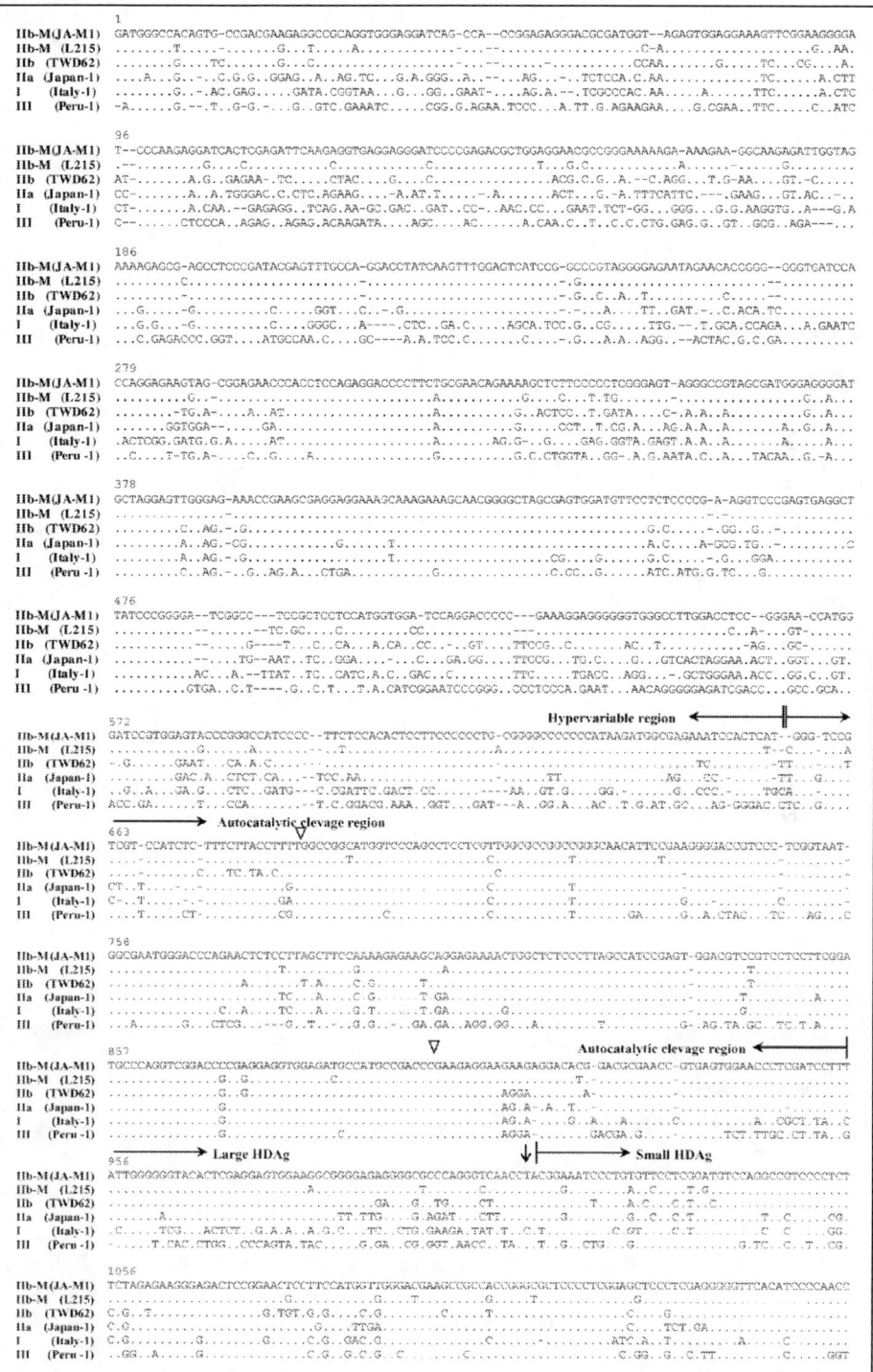

Figure 2. Figure continued on next page. Legend on next page.

```
                1155
IIb-M (JA-M1)   CGCGGGCCGGCTACTCTTCTTTCCCTTCTCTCGTCTTCTTCGGTTAACTTCCGAAGTTCCTCTTCCTCCTCCCTGCTGAGGACCTTTCCCCCGGAGGAAA
IIb-M  (L215)   ..........................................C.....C...C...........................T.....A.......CT.....T........G.
IIb   (TWD62)   ..........................................C...C..C.T.................T...............TT..C..T..C.....G.
IIa  (Japan-1)  ....................T....TC.........A....C.G.C.T.TG...............T....T..........T......T...C.C...G.
I    (Italy-1)  G.........................................C.....C...C...TG................T.....T..........TT...G..T..C.CC...T.
III   (Peru-1)  .................GT.........T....TCG.....A...CCT..CC.G.C....G...........T......TTG......TG........TCC..CC.

                1255
IIb-M(JA-M1)    GCTGCTTTTTCTTGTTCTCCAGGGCCTTCCTTCTTCGGTGGTGCCCGCCTCTCCTGTTCGGTGAACCCTCCCGGGTGTTTCCTCTTCCTAGGTCCGGIAGTC
IIb-M  (L215)   .....TC............................T..T...................................................................
IIb   (TWD62)   .......C...........................................A.......T...........................................
IIa  (Japan-1)  ......C..................G...........G....A..TC.......TG...........G.T.TT...AGG........C...........
I    (Italy-1)  ......C..................G...................G......A..T.........TG............T........CT.A.AGG........
III   (Peru-1)  .T....C.............................T.................................G.............C.TG.CC.TGGGTT...C...C...........A.

                1355
IIb-M(JA-M1)    GACCTCCATCTGATCCGTCCTGGCTCTCTTCGCCGGGGGAGCTCCCTCCCCATCCTTGTCCTT---TCTTATTATTCCTCGGATGTTCCCCAGCCAGGGA
IIb-M  (L215)   ........................................................................................................G
IIb   (TWD62)   .................................G..GC................C.........G........T.,........................C..T...........G
IIa  (Japan-1)  ...............G..GC............C........G....CC.T..---..G.........GAC.
I    (Italy-1)  T...........G....T.G..C..............C....T.......A....CTT..CG.GA....TT.
III   (Peru-1)  A..........GGTT...TG...C.GC..........C....G..TT.G....C.T.C.----..C.AC.G...AAC..C....T.........G

                1455
IIb-M(JA-M1)    TTGTCCTCCTCGAGTCTCTTGAGATTCTTGTTGAATCTTCCCGGAGCTCCTTCTCGAGTTCCTCCTTTCTTCTTCTTCTTCCATCGATCCACTTCCCGAGTGTCT
IIb-M  (L215)   ...................C................G..C...C.....................................T...........
IIb   (TWD62)   ..A..........A...........T.G...........C...............G...C...........C...............G.............T......TG..
IIa  (Japan-1)  ..T..T........T..TGG....CC..G.............A..T........C..........GCCT..T.C..G.TGT........T..T...G..
I    (Italy-1)  ..T..G.......A.TCT.T......T....C..TG..........G..TC..........TAACT.CT......GG.C..C.....G.T...GA..
III   (Peru-1)  ..C..A........A..T....TC.............GCTCG.......A.........C.TTCG.C.GT.CT.C...T.C..T.C.......GTT....GA...

                1555                          HDAg ◄——————————————► Hypervariable region
IIb-M(JA-M1)    CTTCTCTCCCCCTCC---GGACTCTCCTCGCATCGGACTGGCTCATCTTCGAAGAG-GGCGGACGGTCCCGAGAACTTCTATCTTCCTGCTTA--GAA-G
IIb-M  (L215)   ...............T......GT.....................TAG.............G..CT.......T...........,.
IIb   (TWD62)   ...........GG.......QA...G...........G...TTGCGA............T..........CT....
IIa  (Japan-1)  .C..C..TGTT.C..TTC.TC..........TC......T.........TC.AGGTC.-...A-G......TC..TT..C-.C.A...T.T...TTT...A.
I    (Italy-1)  ............T..CG.----..TTCT......A..C...........GCT..A.....-..A...TC..T...CT.....CTT.T..GT--A...
III   (Peru-1)  .C.......T..T..G---A.GTCAG...T..GA...TT..........C.GAG.CC.G..A.C-----.T.GAC..T..CT.......CT.CTAAGG--AGG-A

                1646
IIb-M(JA-M1)    AGGAGTCTCTGGGACGCCTCCGCCC-ACT-CGG.
IIb-M  (L215)   ...........CT......T.....T..-G..-.A.
IIb   (TWD62)   ..........CT.......C.GC..--GG.-....
IIa  (Japan-1)  .....A..GCT....CAAA.....-G.AC-...
I    (Italy-1)  ....A..GCT.....T.GC.GC..GAG.C-.AAG
III   (Peru-1)  ....-....C.A.......C..C.GGCT.CT....
```

Figure 2. Continued. Nucleotide alignment of whole genomic sequences of HDV isolates. The nucleotides are numbered according to Wang et al. The genomic autocatalytic cleavage site (nt. 685/686) is indicated by the white arrowhead. The site (nt. 900/901) corresponding to the cleavage site of the antigenomic RNA is indicated by the black arrowhead. The RNA editing site (nt. 1012) is indicated by an arrow. Sources of isolates are as follows: L215 (AB088679), TWD62 (AF018077), Japan-1 (X60193), Italy-1 (X04451), Peru-1 (L22063). JA-M1(AF309420).

and replication in vivo. Thus, a detailed comparative analysis of HDV genomes between genotype IIb and IIb-M provided a unique opportunity to define the critical genetic features of HDV which determine liver injury. As described later, HDV genotype IIb-M has specific genetic structures in the RNA editing site and the packaging signal sequence of HDAg which could potentially influence the efficiency of HDV replication.[12-17] The observed correlation between HDV genetic structure and clinical characteristics suggests a critical role of variations in the RNA editing site and packaging signal of the HDAg gene in determining the diversity of clinical outcome, even among patients infected with the same genotype of HDV.

Virological Significance of HDV Genotype

Among different HDV genotypes, the difference is highest in the hypervariable region (nt 1598-657) and moderately high in HDAg (nt 957-1597), whereas the autocatalytic regions coding ribozyme activity are well conserved[18] (Fig. 2, Table 1). The hypervariable region is markedly variable even within the same genotype, supporting the notion that this region does not have any relevant biological function aside from the formation of the rod structure of HDV-RNA required for RNA synthesis by RNA polymerase II.[19] On the other hand, the requirement for strict secondary or tertiary structure of the autocatalytic domain seems to be so crucial for full activity of ribozyme needed for rolling-circle mechanism of HDV replication

Table 1. Nucleotide identities among HDV genotypes

	Genotype				
	IIb-M (L215)	IIb (TWD62)	IIa (Japan-1)	I (Italy-1)	III (Peru-1)
Complete sequence (%)	93.8	87.3	78.1	72.7	64.6
Autocleavage region (nt 658-956)	95.6	93.0	92.9	89.2	74.8
Delta antigen (nt 957-1597)	94.4	90.1	81.5	77.9	70.7
Hypervariable region (nt 1598-657)	92.5	82.5	69.0	61.1	54.7

Notes: Numbers given are the nucleotide identities (%) between the isolate JA-M1 (AF309420) and each isolate listed below. Sources of the isolates: L215 (AB088679), TWD62 (AF018077), Japan-1 (X60193), Italy-1 (X04451), Peru-1 (L22063)

that divergence of this region cannot exist among isolates. Therefore, the HDV genetic region other than the hypervariable region or the autocatalytic domain, i.e., the HDAg coding region, confers the clinical difference due to HDV genotypes including IIb and IIb-M.

In the HDAg coding region, the most prominent differences are found in the RNA editing site (Fig. 2) and the packaging signal in the C-terminal region of the large HDAg (Fig. 3, Table 2).[19] Although the coiled-coil domain[20] also shows modest differences, the leucine zipper motif[21] is preserved, and the nuclear localization signal[22] and RNA binding domain[23] are identical in IIb and IIb-M, indicating that these regions are not responsible for genotype-specific liver damages.

Large HDAg, which has 19 additional amino acids (packaging signal sequence) in the carboxyl terminal region of small HDAg, is produced in the late stage of infection through RNA editing of the umber stop codon (UAG) to tryptophan codon (UGG) of the small HDAg

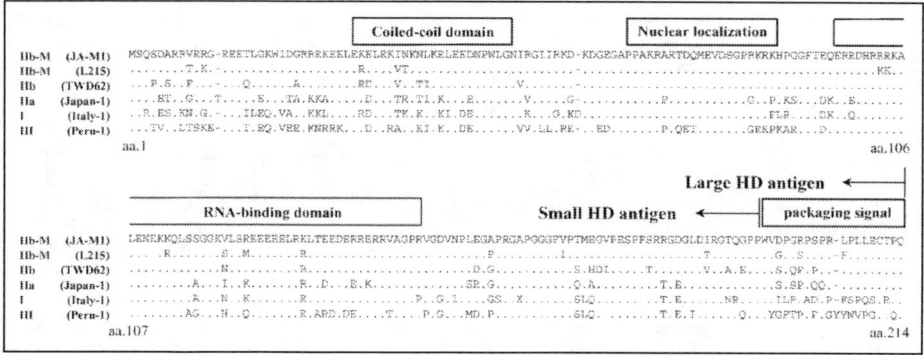

Figure 3. HDAg amino acid alignment of HDV isolates. The amino acids are numbered according to Wang et al. Dots indicate conserved amino acids and dashes indicate missing amino acids. Sources of isolates are as follows: L215 (AB088679), TWD62 (AF018077), Japan-1 (X60193), Italy-1 (X04451), Peru-1 (L22063). JA-M1 (AF309420).

Table 2. *Amino acid identities of HDAg among HDV genotypes*

	Genotype				
	IIb-M (L215)	IIb (TWD62)	IIa (Japan-1)	I (Italy-1)	III (Peru-1)
Complete sequence (%)	92	87	79	70	61
Coiled-coil domain (aa 31-52)	86	77	68	59	64
Nuclear localization domain (aa 68-88)	100	100	86	95	62
RNA-binding domain (aa 97-146)	88	96	84	90	78
Packaging signal domain (aa 195-214)	84	79	74	26	21

Notes: Numbers given are the amino acid identities of HDAg (%) between the isolate JA-M1 (AF309420) and each isolate listed below. Sources of the isolates: L215 (AB088679), TWD62 (AF018077), Japan-1 (X60193), Italy-1 (X04451), Peru-1 (L22063)

gene by host adenosine deaminase.[19] Large HDAg suppresses HDV-RNA replication and promotes virion assembly by extra nuclear export of the HDAg-RNA complex and binding to HBsAg. There are 4 characteristic amino acid differences (codon 198, 200, 201 and 203) in the packaging signal sequence of large HDAg among different genotypes (Fig. 3).[6] As mentioned above, addition of this packaging signal reverses the property of HDAg.[19,24] Although the exact molecular mechanism of this phenomenon is not completely understood, the sequence of 19 amino acids is highly genotype specific. In vitro analysis demonstrated that swapping the packaging signal sequence of genotype IIa with that of genotype I HDAg decreases the viral replication of genotype I, while increasing the replication of genotype II, indicating that this region directly regulates HDV-RNA replication.[13] Thus, the structural characteristics of this region can profoundly influence viral replication. In particular, 2 of the 4 amino acid differences found in IIb-M are located in the proline residues, which are implicated in the assembly process by extra nuclear export of HDAg-RNA complex (Fig. 4).[6] In fact, in a recent study with cultured cells, mutation of the proline residue in this region attenuated the extra nuclear export of large HDAg.[14] However, these data did not directly prove that the C-terminal

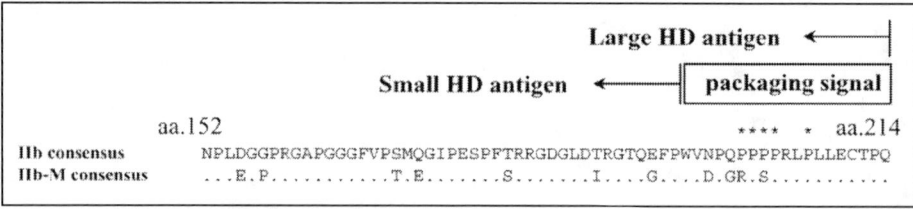

Figure 4. C-terminal end of HDAg of HDV genotype IIb and IIb-M. '*' indicates proline residue in the packaging signal.

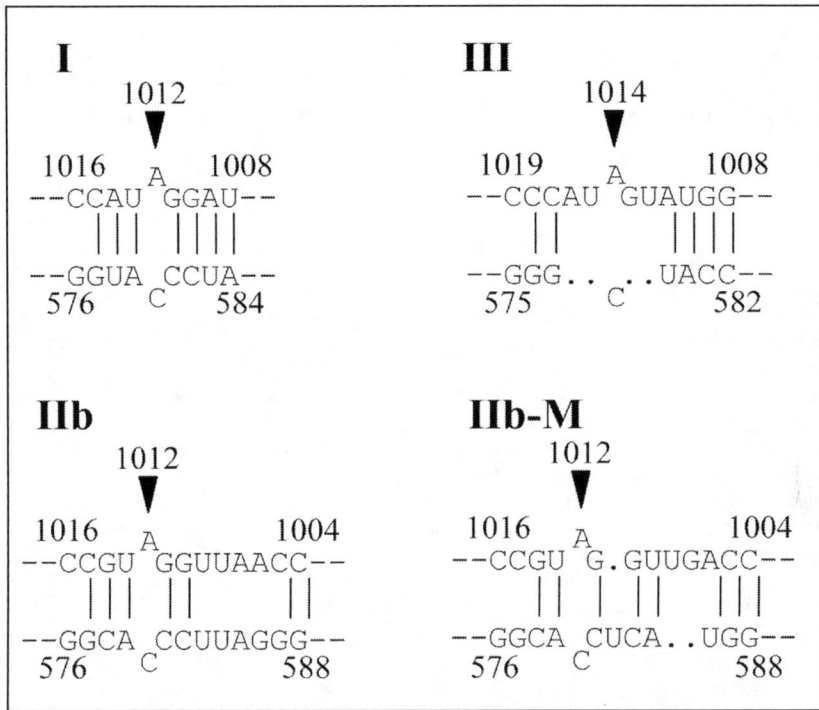

Figure 5. The base-paired structure for RNA editing in genotype I, IIb, IIb-M, and III. Nucleotide sequences of the RNA editing site, which is formed between anti-genome RNA around the edited A residue (nt.1012 for genotype I and II, nt. 1014 for genotype III), and nucleotide sequences of the opposite site of the unbranched rod structure of HDV.

domain structure of HDAg influences the pathogenesis. In the future, in vitro mutational studies should be performed to verify the hypothesis that differences in the packaging signal sequence can modulate HDV replication and lead to progressive disease.

The regulatory mechanism of this RNA editing is not fully understood, but the secondary structure of the antigenomic region corresponding to the 3' end of the small HDAg gene influences the editing efficiency.[12,15,25] A recent in vitro mutational study clearly demonstrated that the base pairing structure surrounding the RNA editing site profoundly influences RNA editing efficiency.[13] In genotype I, the base pairing surrounding this site is particularly strong (4 base pairs on each side), whereas a weaker secondary structure is found within genotype II that is associated with milder liver disease (Fig. 5). In addition, the distinct structure of genotype III is thought to be involved in fulminant hepatitis.[12] Collectively, the specific differences in the base-paired structure of the RNA editing site might explain to some extent the difference in virulence among HDV genotypes. In genotype IIb-M, there is particular disruption of the base pairing structure two base upstream of the editing site, resulting in a characteristic structure in this region distinct from that of genotype IIb.[6] There is a possibility that the unique structure of the RNA editing site of genotype IIb-M may contribute to the observed difference of pathogeneses between genotype IIb and IIb-M.

Conclusion

Further investigation on the relationship between HDV genetic structures and their function and pathogenesis will contribute to a better understanding of HDV biology, and will offer the potential for new therapies for HDV, a disease for which no effective therapy has yet been established.

References

1. Rizzetto M, Canese MG, Gerin JL et al. Transmission of the hepatitis B virus-associated delta antigen to chimpanzees. J Infect Dis 1980; 141(5):590-602.
2. Wang KS, Choo QL, Weiner AJ et al. Structure, sequence and expression of the hepatitis delta (delta) viral genome. Nature 1986; 323(6088):508-14.
3. Makino S, Chang MF, Shieh CK et al. Molecular cloning and sequencing of a human hepatitis delta (delta) virus RNA. Nature 1987; 329(6137):343-6.
4. Casey JL, Brown TL, Colan EJ et al. A genotype of hepatitis D virus that occurs in northern South America. Proc Natl Acad Sci USA 1993; 90(19):9016-20.
5. Wu JC, Chiang TY, Sheen IJ. Characterization and phylogenetic analysis of a novel hepatitis D virus strain discovered by restriction fragment length polymorphism analysis. J Gen Virol 1998; 79(Pt 5):1105-13.
6. Watanabe H, Nagayama K, Enomoto N et al. Chronic hepatitis delta virus infection with genotype IIb variant is correlated with progressive liver disease. J Gen Virol 2003; 84(Pt 12):3275-89.
7. Ivaniushina V, Radjef N, Alexeeva M et al. Hepatitis delta virus genotypes I and II cocirculate in an endemic area of Yakutia, Russia. J Gen Virol 2001; 82(Pt 11):2709-18.
8. Gerin JL, Casey JL, Purcell RH. Hepatitis Delta Virus. In: Blaine Hollinger F, ed. Viral Hepatitis. Philadelphia: Lippincott Williams and Wilkins, 2002:169-182.
9. Wu JC, Choo KB, Chen CM et al. Genotyping of hepatitis D virus by restriction-fragment length polymorphism and relation to outcome of hepatitis D. Lancet 1995; 346(8980):939-41.
10. Sakugawa H, Nakasone H, Nakayoshi T et al. Hepatitis delta virus genotype IIb predominates in an endemic area, Okinawa, Japan. J Med Virol 1999; 58(4):366-72.
11. Nakasone H, Sakugawa H, Shokita H et al. Prevalence and clinical features of hepatitis delta virus infection in the Miyako Islands, Okinawa, Japan. J Gastroenterol 1998; 33(6):850-4.
12. Casey JL. RNA editing in hepatitis delta virus genotype III requires a branched double-hairpin RNA structure. J Virol 2002; 76(15):7385-97.
13. Hsu SC, Syu WJ, Sheen IJ et al. Varied assembly and RNA editing efficiencies between genotypes I and II hepatitis D virus and their implications. Hepatology 2002; 35(3):665-72.
14. Lee CH, Chang SC, Wu CH et al. A novel chromosome region maintenance 1-independent nuclear export signal of the large form of hepatitis delta antigen that is required for the viral assembly. J Biol Chem 2001; 276(11):8142-8.
15. Wu TT, Bichko VV, Ryu WS et al. Hepatitis delta virus mutant: Effect on RNA editing. J Virol 1995; 69(11):7226-31.
16. Yang A, Papaioannou C, Hadzyannis S et al. Base changes at positions 1014 and 578 of delta virus RNA in Greek isolates maintain base pair in rod conformation with efficient RNA editing. J Med Virol 1995; 47(2):113-9.
17. Lee CZ, Chen PJ, Chen DS. Large hepatitis delta antigen in packaging and replication inhibition: Role of the carboxyl-terminal 19 amino acids and amino-terminal sequences. J Virol 1995; 69(9):5332-6.
18. Wu HN, Lai MM. Reversible cleavage and ligation of hepatitis delta virus RNA. Science 1989; 243(4891):652-4.
19. Modahl LE, Lai MM. Hepatitis delta virus: The molecular basis of laboratory diagnosis. Crit Rev Clin Lab Sci 2000; 37(1):45-92.
20. Wang JG, Lemon SM. Hepatitis delta virus antigen forms dimers and multimeric complexes in vivo. J Virol 1993; 67(1):446-54.
21. Chen PJ, Chang FL, Wang CJ et al. Functional study of hepatitis delta virus large antigen in packaging and replication inhibition: Role of the amino-terminal leucine zipper. J Virol 1992; 66(5):2853-9.

22. Xia YP, Yeh CT, Ou JH et al. Characterization of nuclear targeting signal of hepatitis delta antigen: Nuclear transport as a protein complex. J Virol 1992; 66(2):914-21.
23. Lin JH, Chang MF, Baker SC et al. Characterization of hepatitis delta antigen: Specific binding to hepatitis delta virus RNA. J Virol 1990; 64(9):4051-8.
24. Chang MF, Sun CY, Chen CJ et al. Functional motifs of delta antigen essential for RNA binding and replication of hepatitis delta virus. J Virol 1993; 67(5):2529-36.
25. Casey JL, Bergmann KF, Brown TL et al. Structural requirements for RNA editing in hepatitis delta virus: Evidence for a uridine-to-cytidine editing mechanism. Proc Natl Acad Sci USA 1992; 89(15):7149-53.

CHAPTER 2

Hepatitis Delta Virus:
HDV-HBV Interactions

Camille Sureau*

Abstract

The hepatitis delta virus (HDV) is a subviral agent that utilizes the envelope proteins of the hepatitis B virus (HBV) for cell to cell propagation. In infected human hepatocytes, the HDV RNA genome can replicate and associate with multiple copies of the delta protein to assemble a ribonucleoprotein (RNP). However the RNP cannot exit the cell because of the lack of an export system. This is provided by the HBV envelope proteins, which are capable of budding at an internal cellular membrane to assemble mature HDV virions when RNPs are present. This review covers advances in the molecular aspects of the HDV-HBV interactions, with an emphasis on the HBV properties that are instrumental in HDV maturation, in particular the central role of the small HBV envelope protein.

Introduction

Since the initial description of the hepatitis delta antigen (HDAg) in 1977 by M. Rizzetto, the viral agent that was later referred to as the "Hepatitis Delta Virus" (HDV), has been clearly related to the Hepatitis B virus (HBV). HDAg was first observed as a new nuclear antigen present only in liver cells of HBV chronic carriers. It was shown to be associated with a viral agent that was transmissible to chimpanzee in the presence of HBV, and at that time, it was considered a defective virus because of its absolute requirement for HBV coinfection. Soon thereafter the HBV helper functions appeared to be limited to providing the protein content of the delta particle envelope, whereas the inner core was found to contain a small RNA molecule bound to HDAg-proteins. The cloning of the HDV-associated RNA was achieved in 1986, and the sequencing analysis revealed a genome structure that was unique among animal viruses: it was a circular, single-stranded RNA of negative polarity, with an open reading frame coding for the HDAg protein (the only protein that HDV RNA is known to encode), but it lacked the coding capacity for envelope proteins. Moreover, its sequence presented no homology to that of the HBV genome.[1-3]

*Camille Sureau—CNRS, Laboratoire de Virologie Moléculaire, INSERM U76, Institute National de la Transfusion Sanguine, 6 Rue Alexandre-Cabanel, 75739 Paris, France. Email: csureau@ints.fr

Hepatitis Delta Virus, edited by Hiroshi Handa and Yuki Yamaguchi.
©2006 Landes Bioscience and Springer Science+Business Media.

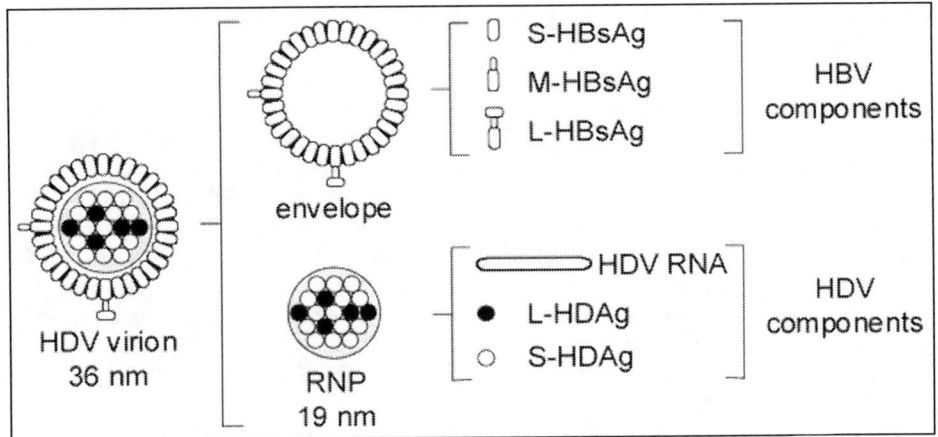

Figure 1. Schematic representation of the HDV particles found in an infectious serum. The virion comprises two types of elements: i) the viral envelope of HBV origin, including the HBV envelope proteins S-HBsAg, M-HBsAg and L-HBsAg, and ii) the ribonucleoprotein (RNP) that comprises the circular genomic RNA associated to multiple copies of the HDV-encoded delta proteins, S-HDAg and L-HDAg.

The Structure of the HDV Particle

The HDV virions are heterogeneous in size with an average diameter of 36 nm (Fig. 1). They display a chimeric structure: the protein content of the envelope consists of the HBV surface proteins, whereas the inner ribonucleoprotein (RNP) complex is made of HDV-specific elements. The HDV RNP is composed of a 1,700 nucleotides single-stranded genomic RNA associated to multiple copies of the HDV encoded delta protein, with a molar ratio of protein to RNA of approximately 200. The delta protein bears the hepatitis delta antigen (HDAg), and it appears under two isoforms: the small form (S-HDAg) is a 195-amino acid polypeptide; and the large form (L-HDAg) is 19 amino acids longer, and it arises as a consequence of a RNA editing event on a replication intermediate of the genome.[3-5]

HDV RNPs can be examined by electron microscopy after subjecting virions to a non-ionic detergent and a reducing agent. They appear as spherical, corelike structures, with no apparent icosahedral symmetry and a diameter of approximately 19 nm. The HDV envelope appears undistinguishable from that of HBV. It consists of a lipid membrane in which the three HBV coat proteins carrying the hepatitis B surface antigen (HBsAg), and designated small (S-HBsAg), middle (M-HBsAg) and large (L-HBsAg), are embedded. These proteins are present at a ratio estimated at 95/5/1, respectively, as opposed to 4/1/1 in the envelope of HBV virions (Dane particles).[1,6]

Therefore HDV appears to be directly dependant on HBV for acquiring its envelope and, for that reason, it should be considered as a defective virus and a satellite of the latter. It does not fulfill the criteria for the definition of a virus; nonetheless, it is referred to as *Hepatitis delta virus* (HDV) for practical reason, and it constitutes the only species of the *Deltavirus* genus. In an infectious serum, HDV particles are present at high titer (up to 10^{11} particles per milliliter) along with different types of HBV particles: the infectious Dane particles and the empty sub-viral particles. They have a similar outer envelope consisting of the HBV coat proteins.[1,6]

Are HBV Helper Functions Limited to Supplying Envelope Proteins to HDV?

Based on the structure of the HDV virion, it would appear that HBV is just a provider of coat proteins. This has been demonstrated experimentally: the transfection of mammalian cells with a cloned HDV cDNA, in the absence of HBV, leads to replication of the HDV RNA and formation of RNPs, demonstrating that HBV is not essential at this level of the HDV cycle. However, the absence of viral RNA or HDAg proteins in the culture medium of transfected cells indicates that progeny RNPs can not be released. Thus, unlike nonenveloped viruses, which are capable of inducing cell membrane lysis for particle release, and contrary to retroviral nucleocapsids that are capable of budding at the plasma membrane in the absence of envelope proteins, HDV cannot release its RNPs by lack of an export system. When cells are cotransfected with HDV cDNA and a plasmid driving the expression of the HBV envelope proteins (in the absence of HBV replication), the RNPs are released as enveloped virions. HBV assists HDV assembly and secretion by just providing the envelope proteins necessary to assemble export vesicles. Moreover, the L-HBsAg and M-HBsAg are not required for this process.[7]

Why Is HBV Best Suited for Assisting HDV?

HBV belongs to the *Hepadnaviridae* family. It is an enveloped virus that contains a circular partially double-stranded DNA, but it replicates its genome through a reverse transcription mechanism. Most important to the HDV-HBV interaction is the type of assembly mechanism that HBV has developed: budding occurs at the membrane of the intermediate compartment (IC), and is driven by the viral coat proteins. The three envelope proteins, S-HBsAg, M-HBsAg and L-HBsAg are found at the surface of HBV virions but in reality, S-HBsAg alone provides the driving force of the budding process. As a result, the vast majority of S-HBsAg proteins are secreted as empty subviral lipoprotein particles.[6]

HBV envelope proteins are coded by a single open reading frame on the HBV genome, using three different sites for initiation of translation. The S-HBsAg protein is 226 amino acid residues in length, and the M-HBsAg protein consists of the S polypeptide and 55 additional residues (the preS2 domain) at the N-terminus. L-HBsAg comprises the entire M polypeptide with an additional N-terminal polypeptide (referred to as the preS1 domain) of 108 residues (Fig. 2).

S-HBsAg is an integral protein, synthesized at the endoplasmic reticulum (ER) membrane (Fig. 3). It is anchored in the lipid bilayer through a N-terminal transmembrane domain (TMD1) between residues 4 and 28. It comprises: the TMD1; a downstream cytosolic loop (residues 29-79); a second TMD (TMD2) between residues 80 and 100; the antigenic loop (residues 101-164) that contains the immunodominant epitopes facing the ER lumen (or the surface of extracellular particles); and a hydrophobic C-terminus (residues 165-226), whose structure is predicted to contain two alpha-helices. M-HBsAg has a membrane topology similar to that of S-HBsAg, with its N-terminal preS2 facing the ER lumen. L-HBsAg, whose C-terminal half consists of the entire S domain, adopts two topologies at the ER membrane: its N-terminal preS domain (preS1 + preS2) is either lumenal (external on secreted virions) or cytosolic (internal on secreted virions). The internal conformation is involved in recruiting the HBV nucleocapsid for virion assembly, and the external form corresponds to a receptor binding function necessary for viral entry (Fig. 2).[6]

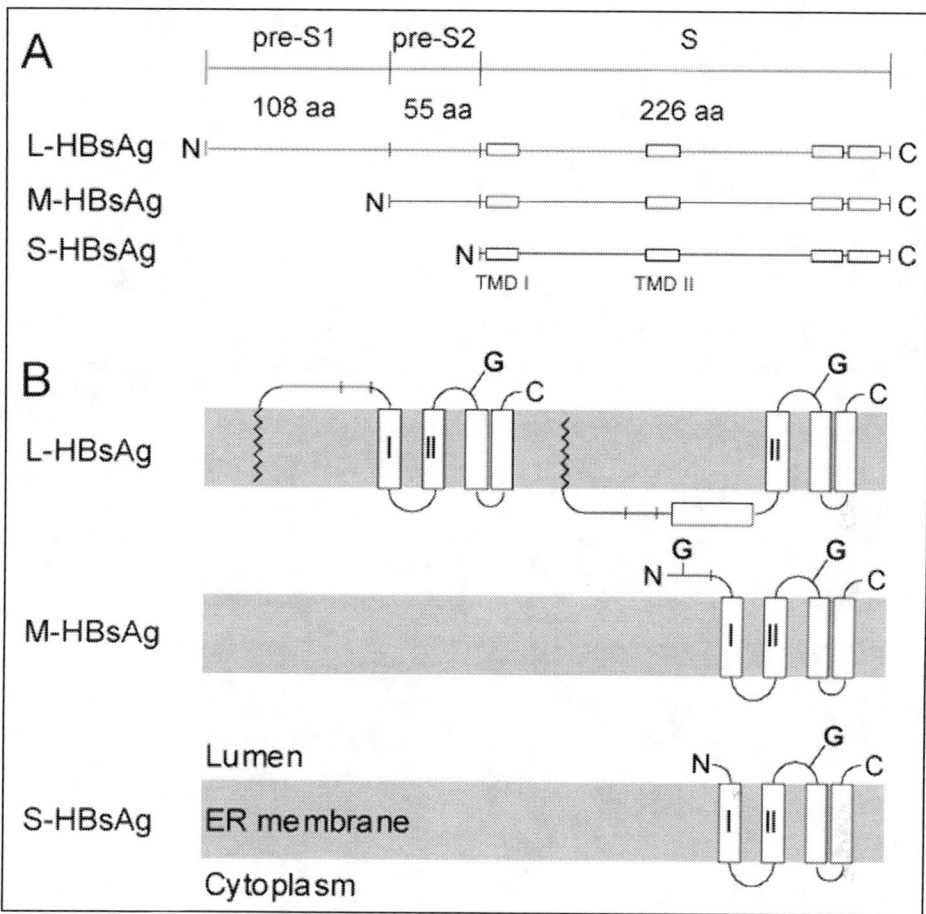

Figure 2. A) Domains of the HBV envelope proteins open reading frame (upper line). The L-HBsAg, M-HBsAg and S-HBsAg proteins are translated from three in-frame initiation sites located at the N-terminus of the preS1, preS2 and S domains, respectively. Open rectangles indicate xperimentally defined TMDs, TMD I, TMD II. B) Membrane topology of the HBV envelope proteins. M-HBsAg and S-HBsAg adopt a similar topology at the ER membrane. The two transmembrane topologies of L-HDAg are represented: the preS domain (preS1 + preS2) can reside on the cytoplasmic side of the ER membrane (right panel), or it can be translocated in the ER lumen (left panel). Broken line indicates the myristate group linked to the N-terminus of L-HBsAg. Grey rectangles represent the lipid ER membrane. Open rectangles represent TMDs. G, glycosylation site.

S-HBsAg proteins can dimerize at the ER membrane through lateral protein-protein interactions, and the resulting aggregates bud spontaneously into the lumen of the IC compartment (Fig. 4). This mechanism is very efficient and central to the maturation of HBV and HDV virions. It is only when L-HBsAg proteins are included in the S-HBsAg aggregates that HBV nucleocapsid can be incorporated in the budding process to assemble mature HBV virions. However this is scarcely observed in comparison with subviral particle formation. It is estimated that more than 99% of HBV-related particles are subviral, and this translates into an

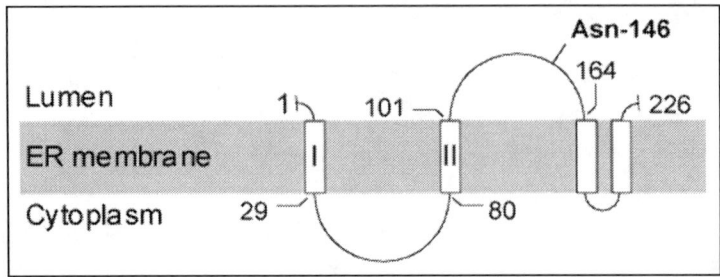

Figure 3. The small HBV envelope protein (S-HBsAg) is synthesized at the ER membrane. It comprises TMD1 (I), amino acid residues 4-28; a cytosolic domain, 29-79; TMD2 (II), 80-100; the antigenic loop, 101-164; two C-terminal TMDs, 165-226. The glycosylation site, Asn-146, in the antigenic loop is indicated.

average infectious serum containing approximately 10^{12} to 10^{13} subviral particles and only 10^8 HBV virions.[6] Thus, HBV appears to have developed an overactive budding mechanism. It is carried out by the S-HBsAg protein itself, and it leads to the production of lipoprotein export vesicles that, for the vast majority, travel empty. It constitutes an export system at the disposal of coinfecting HDV.[7]

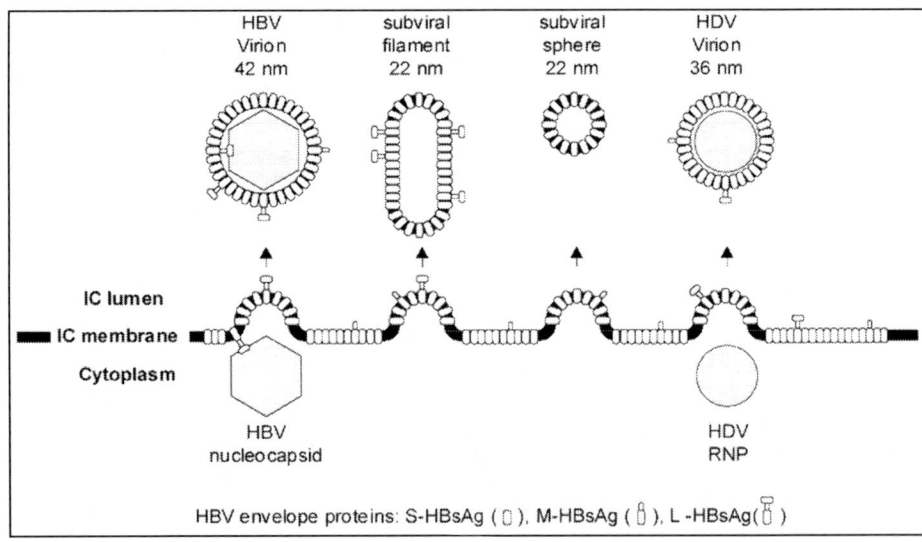

Figure 4. Model for budding of HBV and HDV particles into the lumen of the cellular intermediate compartment (IC). HBV envelope protein aggregates bud spontaneously at the IC membrane. When aggregates include S-HBsAg proteins only, or S-HBsAg + M-HBsAg, it leads to the formation of spherical subviral particles. When aggregates include L-HBsAg in the absence of HBV nucleocapsid, it leads to the secretion of filamentous subviral particles. When aggregates include L-HBsAg in the presence of HBV nucleocapsid, budding leads to the secretion of HBV virions (Dane particles). When HDV RNP is present, it can be included in the aggregates, irrespective of the presence of L-HBsAg. Incorporation of the L-HDAg protein in the HDV envelope confers infectivity.

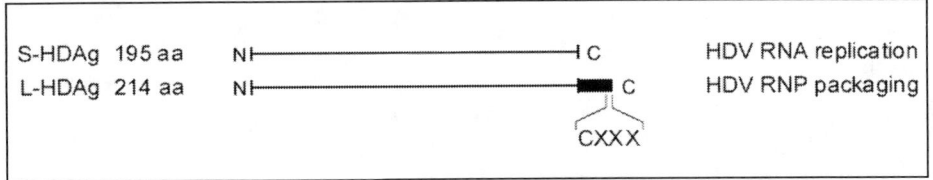

Figure 5. The small form of the delta protein (S-HDAg) is 195-amino acids in length, and the large form (L-HDAg) has a C-terminal extension of 19 amino acids (black rectangle). S-HDAg is required for HDV RNA replication, and L-HDAg is required for packaging of the RNP by the HBV envelope proteins. The signal for farnesylation, CXXX box (C = cysteine, X = any amino acid) is indicated.

How Do S-HBsAg and the RNP Interact with Each Other for HDV Assembly?

In general, virus assembly occurs at a specific site in the cell, and it requires the colocalization of the structural components. For most viruses, newly synthesized proteins are prevented from initiating budding reactions until all virion components are present at the site of assembly. This does not apply to HBV since budding can occur in the absence of HBV nucleocapsid, and this is to the advantage of HDV.

Overall, the formation of progeny HDV virions involves two processes: (i) assembly of the RNP and (ii) formation of the viral envelope. They are directed by two distinct viral species; they are independent of each other and spatially separated, occurring in the nucleus and at the IC membrane, respectively.

Therefore, a critical function that HDV must assume is the targeting of the RNP to the HBV budding system. HDV RNP is composed of the two forms of delta proteins: S-HDAg and L-HDAg (Fig. 5). Both proteins can interact with each other and the genomic RNA to assemble the HDV RNP complex. It is localized primarily in the nucleus, and it must travel to the cytoplasm for packaging with the envelope proteins. This is determined, in part, by the C-terminal 19 amino-acid polypeptide of L-HDAg, which contains a CXXX signal (where C = cysteine and X = any amino acid) for farnesylation.[8] The farnesyl group probably serves to anchor the RNP in the ER membrane where the envelope proteins assemble, and, as expected, treatment of HDV producing cells with a farnesyl transferase inhibitor prevents assembly of RNPs into enveloped particles. When expressed with the HBV envelope proteins in the absence of HDV RNA and S-HDAg, L-HDAg protein can be packaged and secreted in the subviral particles (Fig. 6). The C-terminal 19 amino acids, including the CXXX box are likely to constitute the packaging signal on L-HDAg, since their appending to the C-terminus of a foreign protein, namely c-H-ras, leads to the cosecretion of the latter with HBV subviral particles. Yet, for HDV maturation, it is unclear whether L-HDAg binds to S-HBsAg during budding or beforehand, and whether the free form of L-HDAg associates with the envelope in addition to the RNP-associated form.[2]

Regarding the role of the S-HBV protein, it is worth noting that the member of the *Hepadnaviridae* family closest to HBV, the *Woodchuck hepatitis virus* (WHV), can assist HDV propagation because its S envelope protein (S-WHsAg) is competent for HDV RNP envelopment; whereas the envelope protein of a more distantly related *Hepadnaviridae*, namely the *Duck hepatitis B virus* (DHBV), is unable to fulfill this function. Thus determinants that are specific for HDV maturation on the S-HBsAg protein should be present on S-WHsAg and absent on the small DHBV envelope protein (S-DHBsAg). When compared to S-HBsAg or

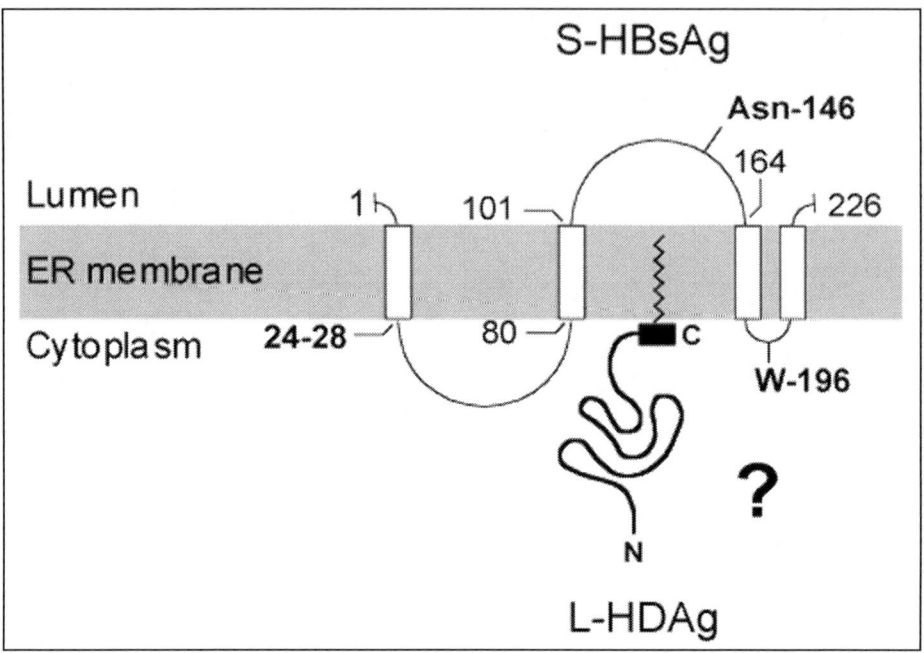

Figure 6. The small HBV envelope protein (S-HBsAg) and the large hepatitis delta protein (L-HDAg) have a crucial role in HDV assembly. They colocalize at the ER or IC membrane and probably interact with each other. The localization of L-HDAg at the ER membrane is thought to depend on the farnesyl group attached to its C-terminus (broken line). Black rectangle indicates the C-terminal packaging signal of L-HDAg. Amino acids of the S-HBsAg protein that are important for envelopment of the HDV RNP are indicated in bold.

S-WHsAg, the S-DHBsAg polypeptide appears to lack the region corresponding to the antigenic-loop (Fig. 3). Indeed, when part of this domain (from residues 107 to 147) is experimentally deleted on S-HBsAg, it leads to a reduction in the capacity of the mutant for HDV maturation.[9] Interestingly, the same mutant was fully competent for packaging HDAg-L proteins, suggesting that the hindrance observed in RNP envelopment may reflect a reduced flexibility of the envelope, which could no longer accommodate an RNP, rather than a lack of binding to the RNP. The deficiency in HDV maturation observed with the antigenic-loop-deleted S-HBsAg, or with S-DHBsAg, could also be explained, at least in part, by the absence of a glycosylation site, since the removal of Asn-146 on S-HBsAg was shown to prevent glycosylation and to inhibit HDV particle formation.[10]

The cytosolic loop of S-HBsAg (residues 28 to 80) has a topology at the ER membrane that is prone to interact with the HDV RNP during virion assembly. In fact, a genetic analysis revealed the importance of residues 24-28 in HDV maturation, but the contribution of this sequence to a direct binding to the RNP was not demonstrated. The same study also revealed that sequences 28-47 and 49-59 do not contain any motif essential for HDV morphogenesis. It is worth noting that the very same region of the cytosolic loop (residues 33 to 59) contains motifs essential for HBV nucleocapsid envelopment.

When the C-terminus of S-HBsAg was examined, it was found that the tryptophane residue at position 196 was important for HDV assembly. This region (164-226) is highly hydrophobic and predicted to contain two TMDs located at positions 173-193 and 202-222. They bracket a short sequence (194-201), including Trp-196, that presents a low degree of flexibility. Hydrophobicity and secondary structure predictions are compatible with the orientation of Trp-196 at the cytosolic side of the ER membrane in a position potentially adequate for interaction with the RNP.

Two observations that bear on the RNP envelopment mechanism, can be made. First, HDV assembly was not completely abolished when residues 24-28, Asn-146 or Trp-196 were singly deleted on S-HBsAg, suggesting that direct interactions, if any, between HDV RNP and S-HBsAg may involve several domains or residues distributed on the entire polypeptide. They could adopt a spatial proximity conferred by the complex organization of the polypeptide at the ER membrane. Second, the fact that motifs identified as essential to HDV assembly, such as residues 24-28 or Trp-196, are dispensable for subviral particle secretion and yet strictly conserved among HBV and WHV isolates, suggests that the selection pressure that has led to their conservation concerns functions other than those involved in subviral particle assembly. Thus, HDV RNP envelopment may rely upon S-HBsAg residues that are operative in maturation or infectivity of HBV virions. But this remains to be proven experimentally.[11]

For a better understanding of the HDV maturation process, the determinants of incorporation of L-HDAg proteins into the subviral particles need to be sorted from those involved in RNP envelopment. In the former case, assembly should proceed through colocalization of the particle components and a specific interaction between L-HDAg and S-HBsAg; whereas in the latter case, assembly is likely to depend also on the constraints exerted on the envelope to accommodate a 19 nm RNP. The fact that HBV manages to assemble three types of particles, namely the 22 nm subviral spheres, the filamentous forms and the 42 nm Dane particles, indicates that S-HBsAg can modulate its intrinsic membrane bending force. Therefore, the flexibility of the HBV envelope protein is another important characteristic of this virus (Fig. 4).

But is there an actual need for a direct interaction between S-HBsAg and L-HDAg, or could HDV RNPs be passively incorporated in the HBV budding vesicles? In the light of a recent study that measured the concentration of HDAg proteins at up to six millions copies per infected cell, it seems that the encounter between L-HDAg and S-HBsAg, or any cellular factor involved in this process, should be facilitated.[4] It would thus be interesting to test the capacity of other viruses such as *Coronaviruses* or *Spumaviruses* to substitute for HBV in HDV assembly. For both viruses, as for HBV, budding occurs at the post-ER/preGolgi membrane, and it is driven by the viral coat proteins.

However, one could argue for the requirement of a specific interaction between S-HBsAg and HDV RNPs based on the following observations: (i) sera collected at the peak of an acute HDV infection usually contain a high proportion of HDV virions to empty subviral particles, (ii) although produced less abundantly than S-HBsAg, S-DHBsAg proteins expressed in human cells are not capable of HDV maturation, (iii) a direct protein-protein interaction between S-HBsAg and L-HDAg has been reported using a far-Western binding assay, and (iv) synthetic peptides specific for HBV envelope proteins have been shown to bind both L-HDAg and S-HBsAg proteins.[1-3]

The Infectivity of the HDV Virions

For propagation, secreted virions must be redirected to a non infected cell. Therefore, it was expected for the L-HBsAg protein, which mediates HBV infectivity, to be required as an integral component of the HDV envelope. This was demonstrated in an in vitro culture system: HDV particles coated with the S-HBsAg protein, or S-HBsAg and M-HBsAg, were not infectious when tested on primary hepatocyte cultures; but when L-HBsAg was coexpressed with S-HBsAg, infectivity was restored.[12]

In an HBV-infected cell, the incorporation of L-HBsAg proteins in budding vesicles has the following consequences: (i) it makes the budding unit competent for HBV nucleocapsid envelopment; (ii) it exerts a partial retention of subviral particle secretion; and (iii) it induces a change in the shape of the empty particles (from spherical to filamentous). It is not clear if these functions are fully operative or partially altered when the HDV RNPs are present (Fig. 4).[6]

Overall, the HDV life cycle depends on only two HBV elements: the S-HBsAg protein for the export of the RNP, and the L-HBsAg protein for entry in a noninfected hepatocyte. As for the M-HBsAg protein, its role needs L-HBsAg to envlope, is not essential for assembly or in vitro infectivity of HDV. But contrary to HBV that its nucleocapsid only in the presence of L-HBsAg, HDV can be coated quite efficiently with an envelope devoid of L-HBsAg. And it is likely that most HDV particles present in an infectious serum are coated with S-HBsAg proteins only (or S-HBsAg and M-HBsAg), and that only a minority contain L-HBsAg. The inclusion of L-HBsAg in the HDV envelope is thought to occur through lateral protein-protein interactions that are established between L-HBsAg and S-HBsAg at the ER membrane before budding, but the proportion of L-HBsAg proteins that are needed in the envelope for infectivity is unknown. With regard to virus entry into the host cell, it seems reasonable to assume that HBV and HDV use the same cellular receptors on the human hepatocyte, but the identities of the receptors remain unknown. At the post-binding steps, internalization of an HDV RNP or that of an HBV nucleocapsid, most likely follows a specific pathway to which S-HBsAg and L-HBsAg may participate.[1,6,12]

What Are the Effects of HDV Infection on the HBV Life Cycle?

Since HDV is directly dependent on HBV for propagation, it can be transmitted concomitantly with HBV to an individual who has no history of prior HBV infection—this is referred to as the coinfection pattern—or it can be transmitted to an HBV chronic carrier—this is referred to as the superinfection. Coinfections are often acute and self-limited, and they are characterized by a concomitant replication of both HBV and HDV; whereas superinfections often cause severe acute hepatitis and chronic type D hepatitis in 70% of the cases.[1] They also lead to the inhibition of HBV replication during the acute phase of HDV infection. This phenomenon has been described in both humans and the experimentally infected chimpanzee, but it remains poorly understood. It could result from a direct suppression of HBV replication exerted by the coexpressed HDV proteins, RNA or RNPs, or it could be the consequence of an indirect interfering mechanism which may involve inflammatory cytokines.[1]

There are at least three HDV genotypes with different geographic distributions and a sequence divergence as high as 38% at the nucleotide level. Genotype I is ubiquitous, genotype II has been found primarily in east Asia, and genotype III only in South America. Eight HBV genotypes (designated A to H) have been described, presenting also different geographic distributions. The heterogeneity of the disease pattern caused by HDV-HBV infections, which has been observed worldwide, may depend on specific HDV on HBV genotypes. The most severe form has been recorded in South America where a genotype III HDV was associated to a genotype F HBV.[1]

Conclusion

Among enveloped viruses that achieve maturation through a nucleocapsid-independent assembly mechanism, HBV has developed the most active budding process. It is carried out by the S-HBsAg envelope protein itself, and it leads to the formation of a large excess of empty subviral particles over mature virions. For that reason, HBV appears the best-suited virus for assisting a coatless virus such as HDV to export its genome from an infected cell by supplying the transport vesicles. Though not proven, a specific interaction between the HDV RNP and S-HBsAg is probably needed to ensure an efficient virion assembly. Recent experiments based on genetic analysis have contributed to the understanding of the intracellular HDV-HBV interactions involved in HDV maturation, but it is expected that further progress in this direction will be achieved when ultrastructural data for both the HBV envelope and the HDV RNP become available.

References

1. Gerin JL, L CJ, Purcell RH. Hepatitis Delta Virus. In: Knipe DM, Howley PM, eds. Fields Virology. Philadelphia, PA: Lippincott Williams & Wilkins, 2001:3037-3050.
2. Taylor JM. Replication of human hepatitis delta virus: Recent developments. Trends Microbiol Apr 2003; 11(4):185-190.
3. Lai MM. The molecular biology of hepatitis delta virus. Annu Rev Biochem 1995; 64:259-286.
4. Gudima S, Chang J, Moraleda G et al. Parameters of human hepatitis delta virus genome replication: The quantity, quality, and intracellular distribution of viral proteins and RNA. J Virol 2002; 76(8):3709-3719.
5. Rizzetto M, Hoyer B, Canese MG et al. Delta agent: Association of delta antigen with hepatitis B surface antigen and RNA in serum of delta-infected chimpanzees. Proc Natl Acad Sci USA 1980; 77(10):6124-6128.
6. Ganem D, Schneider R. Hepadnaviridae: The viruses and their replication. In: Knipe DM, Howley PM, eds. Fields Virology. Philadelphia, PA: Lippincott Williams & Wilkins, 2001:2923-2969.
7. Wang CJ, Chen PJ, Wu JC et al. Small-form hepatitis B surface antigen is sufficient to help in the assembly of hepatitis delta virus-like particles. J Virol 1991; 65(12):6630-6636.
8. Glenn JS, Watson JA, Havel CM et al. Identification of a prenylation site in delta virus large antigen. Science 1992; 256(5061):1331-1333.
9. O'Malley B, Lazinski D. A hepatitis B surface antigen mutant that lacks the antigenic loop region can self-assemble and interact with the large hepatitis delta antigen. J Virol Oct 2002; 76(19):10060-10063.
10. Wang CJ, Sung SY, Chen DS et al. N-linked glycosylation of hepatitis B surface antigens is involved but not essential in the assembly of hepatitis delta virus. Virology 1996; 220(1):28-36.
11. Jenna S, Sureau C. Mutations in the carboxyl-terminal domain of the small hepatitis B virus envelope protein impair the assembly of hepatitis delta virus particles. J Virol 1999; 73(4):3351-3358.
12. Sureau C, Guerra B, Lanford RE. Role of the large hepatitis B virus envelope protein in infectivity of the hepatitis delta virion. J Virol 1993; 67(1):366-372.

CHAPTER 3

Structure and Replication of Hepatitis Delta Virus RNA

John M. Taylor*

Abstract

This review focuses on the RNAs of HDV, with emphasis on RNA structure, RNA transcription, and post-transcriptional RNA processing. Included is an evaluation of two current models of HDV RNA replication.

Introduction and Scope

Hepatitis delta virus (HDV) was first discovered in 1977 through the work of Rizzetto and coworkers.[1] Around 1986 several other labs began to work on the molecular virology of this agent and over and over, HDV has provided us with intriguing and unique phenomena in molecular virology. There are still important questions that need to be resolved. However, as a problem of natural infections in humans, HDV is apparently slowly "vanishing".[2]

The molecular biology of the HDV RNAs and their mechanism of replication depend upon the production of two related virus-encoded proteins, the small and large forms of the delta antigen (δAg), referred to here as δAg-S and δAg-L, respectively. The properties of these essential proteins are the focus of Chapter 4. In the present chapter, the focus will be on the RNAs of HDV, with consideration of such features as structure, transcription, post-transcriptional processing, and stabilization. The reader might also want to consider earlier reviews of these topics.[3-10]

The RNAs

Many different complete HDV RNA sequences have now been reported.[11-13] Most sequences are at or about 1,679 nucleotide (nt) in length. We will in this review use the numbering system of Kuo et al.[13] The origin of this numbering is indicated in Figure 1. For the genomic RNA the numbering increases for the 5'-3' direction. For the antigenome, the numbering decreases for the 5'-3' direction.

Consider now the three HDV RNA species that get the most attention. As diagrammed in Figure 1, they are the genome, the antigenome, and the mRNA. The RNA species that is assembled into virus particles is, by definition, the genome. It is a single-stranded RNA with a circular conformation. Within cells where this genome is replicating there is also present typically 5-20 fold lower amounts of the antigenome, an exact complement of the genome. The third

*John M. Taylor—Fox Chase Cancer Center, 7701 Burholme Ave., Philadelphia, Pennsylvania 19111, U.S.A.; Email: jm_taylor@fccc.edu

Hepatitis Delta Virus, edited by Hiroshi Handa and Yuki Yamaguchi.
©2006 Landes Bioscience and Springer Science+Business Media.

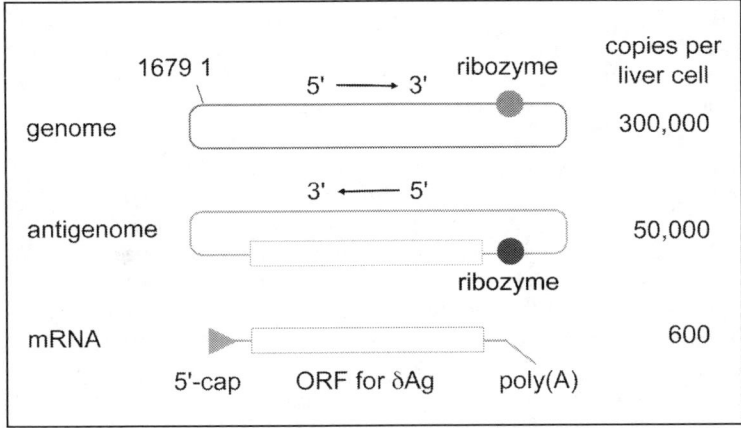

Figure 1. Three processed RNAs of HDV. These are the genome, antigenome and mRNA. Indicated on the genome and antigenome are the positions of ribozyme cleavage. Indicated on the mRNA, with its open reading frame for δAg, are the 5'-cap and 3'-poly(A). The abundances of each species are as reported previously.[16] Indicated is the position of the origin of the 1,679 nucleotide sequence of Kuo et al.[13]

RNA species is of the same polarity as the antigenome. It is linear, 5'-capped[14] and 3'-polyadenylated.[15] Its length of about 800 nt spans the open reading frame for δAg and is considered to be its mRNA. As will be explained in more detail, not just the mRNA but all three of these RNAs have undergone one or more forms of post-transcriptional RNA processing.

Actually, during HDV RNA replication there are minor amounts of yet other processed RNAs. These include relatively low amounts of dimers and even trimers of the unit-length, for both genomic and antigenomic polarity.[16] For these multimers as well as for the monomers, the majority of the RNA is in a circular conformation.

The 5'-end of the mRNA was initially mapped at position 1631 using primer extension assays.[15] In later studies using 5'-RACE procedures, it was mapped to position 1630.[17] This is the predominant 5'-end but there seem to be other sites that are less abundant and less specific.[17] Uncertainty arises because in many of the early studies the mRNA was of low abundance, typically <2% relative to the antigenomic RNA. At the right side of Figure 1 is indicated what has been deduced for the number of the three HDV RNAs per average hepatocyte in an HDV infected liver.[16]

While the unit-length genomic and antigenomic RNAs are primarily in a circular conformation, there are unit-length linear forms and their nature may be complex. Some may be species whose ends have been defined by ribozyme cleavage and have yet to be circularized. Others may be circles that have been reopened by ribozyme cleavage or exonuclease action. In one study, many 5'-ends of linear genomic monomers detected within the liver of an infected animal were mapped to a specific site that was not a site of ribozyme cleavage.[18] As considered later, the site is more likely to be a site of endonucleolytic opening on preformed circles than a site of initiation of transcription.

Yet another class of HDV RNAs must exist. These are the unprocessed and partially processed linear RNAs of both genomic and antigenomic polarity. Some studies have detected species of much greater than unit-length that may be examples of unprocessed nascent transcripts.[19] Presumably because of their transitory nature and/or the techniques used, the unprocessed species are more difficult to detect and characterize. Such RNAs should include species containing the anticipated 5'-ends of nascent transcripts. However, at this time we have no data for the detection of 5'-ends for genomic RNAs. As explained below, the 5'-end of the mRNA species might correspond to an initiation site of antigenomic RNA.

RNA Structure

The first complete sequence of HDV genomic RNA provided not only evidence that (at least some of) this RNA was circular in conformation, but also the computer prediction that the RNA has the ability to fold on itself, into what has become known as "an un-branched rod-like structure".[11] The genomic RNA is predicted to have 74% base pairing, and a negative free energy of 805 kcal.[13] Other studies have used electron microcopy to detect the rod-like structure.[20] Also, using electrophoresis under nondenaturing conditions, the HDV RNA behaves as if it were double-stranded.[21] As will be subsequently considered (and in Chapter 5), the action of RNA-editing also supports the interpretation that at least part of the antigenomic HDV RNA can fold into the rod-like structure. Yet another piece of circumstantial evidence is derived from the observation that the circular RNA, once formed, is not recleaved. One interpretation, is that the ribozyme domain is now inactivated by being forced to adopts the rod-like folding. However, even with all this evidence in support of the potential to fold into such a structure, it must be pointed out that we do not have direct proof that this structure predominates in vivo.

Even though there is a tendency towards ascribing a single rod-like structure to HDV RNAs we must make clear that multiple structures are not only possible but probably also essential. For the case of the plant viroids, there are elegant studies supportive of the concept of metastable RNA structures. For example, with potato spindle tuber viroid (PSTVd) in addition to a predicted rod-like folding there is also a predicted hairpin structure that is not present on that rod.[22] Good genetic evidence shows that this hairpin must be able to form for PSTVd replication to occur. Therefore it should come as no surprise that HDV RNAs also need more than one structure. For example, the folding of the ribozyme domains on both the genome and antigenome are in no ways like that of the rod-like structure. No doubt other examples of metastable states for HDV RNAs will be found.

In addition to the known structural features of HDV RNAs there are also features that at this time can only be inferred from the primary sequence. Specifically, there are patches of G and C that have been noted by Branch and co-workers.[23] No significance for these has been shown as yet, although it has been speculated that they might facilitate some aspect of transcription, such as initiation.

One approach to understanding intra-molecular structure of HDV RNAs, genomic and antigenomic, has been to test cross-linking produced by UV irradiation. Initially, this strategy led to the definition of a genomic RNA site, referred to as an E-loop motif.[24] More recent studies have found a cross-linkable site on antigenomic RNA, but it is not of the same motif.[25] It should be noted that such cross-linking has been previously defined for other RNAs such as the viroid PSTVd and the host cell 5S RNA.[26] However, for both HDV and the viroids, the biological relevance is not proven.

Studies have been reported of the binding to HDV RNA sequences of the protein kinase PKR. These studies were performed in vitro and with less than full-length HDV RNA sequences. It was concluded that a region of the antigenomic RNA folded into a structure other than the rod-like structure, that under certain conditions was needed for PKR binding.[27] At this time, there are data both for and against the relevance of PKR activation for HDV.[27]

One important comment regarding HDV RNA structure and processing, is that the initial structure for the HDV RNAs might be dictated by transcription. Others have referred to this phenomenon as "co-transcriptional folding". Specifically, it is proposed that structures that can be achieved during transcription might actually be different from those detected for the folding-refolding of complete and processed RNAs.[28]

Experimental Systems for the Initiation of HDV Replication

The original experimental studies of HDV replication were carried out using infections of chimpanzees.[29] Soon it was realized that HDV would also replicate in woodchucks, if these animals were coinfected with woodchuck hepatitis virus, WHV, which is very similar to HBV.[30] Later, it was even found that after injection of natural virus, even some of the hepatocytes of a mouse could be infected.[31] More recently, one study implanted human hepatocytes beneath the kidney capsule of a mouse and subsequently showed that the animal could now be infected via an i.v. injection with HDV (or HBV).[32]

Separate studies showed that infection could be achieved for primary cultures of hepatocytes of primate or woodchuck origin.[33,34] Such primary cultures are expensive and difficult to establish. Therefore, it should not be surprising that many labs moved to more convenient systems.

The first simplified system was to construct expression vectors with tandem dimers or trimers of unit-length cDNA clones of HDV sequences. These were then transfected into established cell lines and HDV RNA replication was achieved.[35] Soon it was realized that at early times after transfection, most of the detected unit-length HDV RNA, be it genomic or antigenomic, was DNA-directed rather than RNA-directed. For this reason cDNA constructs were reduced to sizes just larger than unit-length. In this way it was possible to be sure that the accumulation of unit-length transcripts was based on RNA-directed rather than DNA-directed transcription.[21]

Another strategy was to transcribe HDV RNA sequences in vitro, and then transfect these into cells. This has never led to HDV replication. However, if the transfected RNA enters a cell already expressing the δAg-S, then replication can be initiated.[36] Similarly, if the transfected RNA is already in a complex with δAg, then replication can be initiated. This can be a complex between in vitro transcribed RNA and recombinant δAg.[37] However, it also possible to use ribonucleoprotein complexes released from natural virus, or even the intact virus itself.[38]

In 1998, Lai and co-workers showed that these direct sources of δAg could be replaced by an indirect source. They found that when mRNA for δAg was cotransfected along with HDV RNA of greater than unit-length, HDV replication occurred.[39] In this method, δAg-S must be translated before it can contribute to HDV genome replication. This experimental strategy uses the mRNA species in amounts 100-fold more relative to antigenomic than would be expressed in a natural infection. Not surprisingly then, the HDV genome replication is faster and more extensive.

It should also be noted that it is possible to use recombinant HDV sequences to infect animals. The first example was to transfect a HDV cDNA construct into the liver of an HBV infected chimp.[40] This has been repeated with woodchucks.[41] Another variation has been to use a hydrodynamics-based transfection of a mouse with either cDNA or even a combination of greater than unit-length RNA and mRNA.[42]

While many different methods have been developed to study aspects of HDV replication, one has to be concerned that some transfection strategies may give answers very different from what happens in natural infections. One should expect that the stoichiometry of δAg molecules per molecule of RNA template might be an important parameter of HDV replication.[43] It might also be that this antigen might have to be newly formed. For example, we know that this protein can accumulate with time, various levels of phosphorylation,[44] which in turn could change the biological properties of the protein (as discussed in Chapter 4).

Table 1. Partial list of experimental systems for studying HDV genome replication

Strategy for Initiation of Genome Replication	Species / Tissue or Cell	Reference
Infection with virus	human / liver	Rizzetto et al (1977)[1]
	chimpanzee / liver	Chen et al (1986)[16]
	woodchuck / liver	Ponzetto et al (1984)[30]
	mouse / liver	Netter et al (1993)[31]
	chimpanzee / hepatocytes	Sureau et al (1991)[34]
	woodchuck / hepatocytes	Taylor et al (1987)[33]
Transfection with virus	variety of cultured cell lines	Bichko et al (1994)[38]
Transfection with DNA	chimpanzee / liver	Sureau et al (1989)[40]
	woodchuck / liver	Rapicetta et al (1993)[41]
	mouse / liver	Chang et al (2001)[42]
	mouse / skeletal muscle	Polo et al (1995)[100]
	established cell lines	Kuo et al (1989)[35]
Transfection with RNA	mouse / liver	Chang et al (2001)[42]
	established cell lines	Modahl & Lai (1998)[39]

Now the virus particles that are assembled during a natural infection will contain RNA genomes that have accumulated sequence changes. Some of these changes will compromise the replication competence of RNA. For one, there is the ADAR editing (Chapter 5), which can be >50% and leads to HDV RNAs that can no longer achieve synthesis of the essential δAg-S. For another, there are mis-incorporations that arise during RNA-directed RNA replication. The frequent change at nt 1375 might be of this class.[43] A serious consequence of these and other changes is that a large fraction of the assembled genomes are not fully replication competent.

This accumulation of compromising changes arises not only during natural infections but will also apply to HDV particles assembled by cotransfection of cells with plasmids to express the envelope proteins of the helper virus, HBV.

Table 1 summarizes various experimental systems that have been used to study HDV replication. Certainly the choice of experimental strategy used to initiate HDV replication can depend upon the question being asked. However, as we move to questions, such as how particles can attach and enter into susceptible cells, it will become more relevant to use replication-competent virions and naturally occurring susceptible cells.

RNA-Directed Transcription

For some time it has been clear that HDV genome replication does not involve DNA intermediates, and must therefore use RNA-directed transcription.[16] However, the next step of clarifying which polymerase(s) are used for this transcription has remained largely unsolved and even then, controversial. Currently there are three types of evidence for the involvement of the host RNA polymerase II.

The first class of evidence, albeit circumstantial, is that one of the HDV RNAs looks like a pol II transcript. The mRNA has at least three characteristics that provide a good circumstantial argument for pol II involvement. This mRNA has a 5'-cap, a 3'-poly(A), and on the presumptive antigenomic RNA precursor, one can see essential signals for polyadenylation, namely, a AAUAAA signal, 3' of this a CA acceptor site for poly(A) addition, and further 3', a short sequence rich in G and C.[15] These are all features seen for the polyadenylated mRNA of the host cell. (These features are indicated in Fig. 2.)

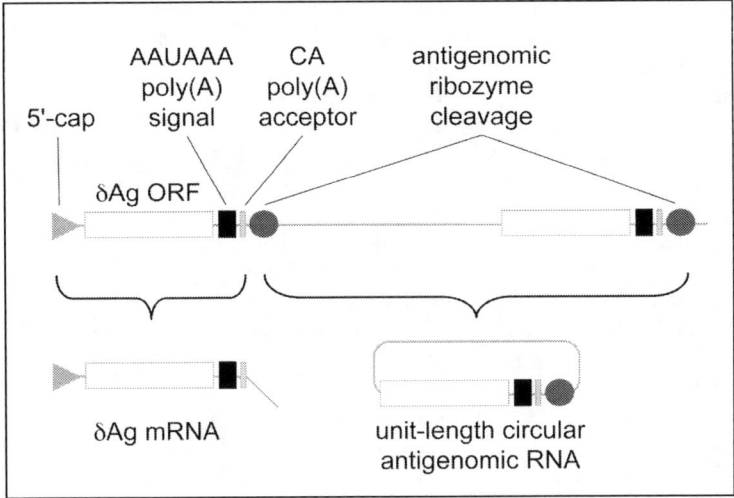

Figure 2. Processing sites on antigenomic RNA. Shown in the upper half is an antigenomic RNA transcript that has initiated at what becomes the 5'-end of the mRNA species. This transcript is shown to have elongated to what would correspond to about 1.5-times around a circular genomic RNA template. Shown in the lower half are two RNAs that could be processed from such a transcript. First is poly(A)-processing to produce the mRNA for δAg. Second, is ribozyme cleavage followed by ligation, to produce unit-length circular antigenomic RNA. As discussed in the text, it is now considered possible that these two processing pathways are largely mutually exclusive.

The second class of evidence is linked up with the use of the polymerase inhibitor α-amanitin. At relatively low doses this inhibitor will specifically block DNA-directed RNA transcription by pol II. Higher doses are needed to block pol III and even higher doses to block pol I.[45] Studies with several different experimental approaches, but in all cases involving amanitin, agree that such low doses can block the transcription and accumulation of both the HDV mRNA and of genomic RNA. (As will be discussed, different results have been reported for the inhibition of new antigenomic RNA.)

A third class of evidence is now being achieved by immuno-precipitation strategies. One lab first showed that δAg could be bound to the HDV RNA by formaldehyde treatment of intact cells in which HDV replication was occurring.[46] Then, in a similar but more difficult experiment, they were able to show that pol II could be cross-linked in vivo to HDV RNAs (Niranjanakumari, Lasda, Brazas, and Garcia-Blanco, personal communication). One can expect that this approach will be used to test whether polymerases other than pol II can be found bound to HDV RNAs. It may also be able to capture other host proteins involved in the transcription and processing of HDV RNAs.

There is a fourth type of evidence that is realized to be necessary. These are in vitro studies in which attempts are made to get nuclear extracts or even better, purified polymerases, to act on added HDV RNAs. Certain positive results were reported but subsequently could not be reproduced.[47] Other studies should be viewed as at best, partially positive. These are studies in which added HDV RNAs were able to act as templates but the transcription was exclusively 3'-end addition and the lengths of the added sequences were very short, typically <50 nucleotides.[14,48,49] In one such study, a 3'-addition was achieved after a site-specific endonucleolytic cut on the HDV RNA produced the relevant 3'-OH site.[49] In another set of studies, δAg was added to this in vitro transcription, and it was shown that longer additions were achieved.[50,51] This effect is considered in more detail in Chapter 6.

However, in all these in vitro studies there has yet to be demonstrated reproducible and extensive transcription for an added HDV RNA template. Also needed is clear evidence for the initiation of transcription, if in fact this can be achieved. In one study it was claimed that initiation was achieved but the data do not clearly demonstrate this.[48]

Now we have to consider a serious complication to our understanding of HDV transcription. Two studies have reported that the transcription from a genomic RNA template to make antigenomic RNA involves a polymerase other than pol II.[19,52] These data are based solely on what is interpreted as a resistance of the transcription to high doses of amanitin; for DNA-directed RNA transcription in animal cells this is usually an indication of transcription by pol I.[45] It should be noted that such antigenomic RNAs are significantly less abundant than genomic RNA, and thus more difficult to detect. From these reports the authors have proposed that pol I is involved, and they have presented a model in which HDV replication involves the incoming genomic RNA acting first as a template for pol II to make a mRNA precursor and then for this same RNA, to act as a template for the non-pol II enzyme to transcribe multimers of antigenomic RNA (Fig. 4). This model is discussed later.

In studies of HDV it has often been valuable to consider the analogy between HDV and the plant viroids.[8] Just one aspect of the analogy is that these agents, like HDV, are totally dependent on parasitizing a host RNA polymerase to achieve transcription of their genomes. This being said, can we gain insights for HDV, by considering which host polymerases are used to transcribe viroid RNAs? Two different polymerases have been implicated. One class of viroids, of which potato spindle tuber viroid (PSTVd) is the prototype, are considered to be transcribed by the plant pol II.[53] The second class, with avocado sunblotch viroid (ASBVd) as the prototype, are considered to use a nuclear-encoded RNA polymerase that normally acts in the plant chloroplast.[54]

In the above discussion of HDV and viroid transcription it is accepted that a host RNA polymerase that normally uses DNA as a template can be redirected to carry out transcription on an RNA template. If this is indeed true, we still need to understand what it is that makes these RNAs able to so achieve the redirection. Some authors have tried to define what might be a promoter element on the HDV RNAs.[48] Most studies have accepted the possibility that the 5'-end of the mRNA could well be a unique initiation site, predominantly at position 1630[17] on the 1679 nt sequence of Kuo et al.[13]

Some comment needs to be made regarding the possible existence of a mammalian RNA polymerase that does not need to be redirected. Several years ago an RNA-directed RNA polymerase was purified from a plant.[55] Soon after this polymerase was finally cloned, it was realized that all plants contain at least one copy of such a gene.[56] Next it was realized that this polymerase can play an essential role in the amplification of small interfering RNAs (siRNA).[57] However, there is as yet no evidence that such a polymerase is involved in viroid replication. In addition, in mammalian genomes a sequence for such a polymerase has not yet been found nor has such an activity been detected.

Template-Switching and Recombination

For some positive-strand RNA viruses it is known that during replication there can be an inter-molecular template-switching.[58] This is a form of recombination. In contrast, such recombination has not been found for negative-strand viruses. HDV is essentially negative-stranded since the only open-reading frame is on the antigenome. An unsuccessful attempt was made to detect such recombination between genetically marked HDV RNAs replicating in cultured cells (Wu and JMT, unpublished observations). As an alternative approach, another group studied HDV RNAs in patients infected with more than one form (genotype) of HDV. Their evaluation of the RNA sequences present in such patients is that intermolecular recombination could have occurred.[59] It is worth noting that claims have been made for intermolecular

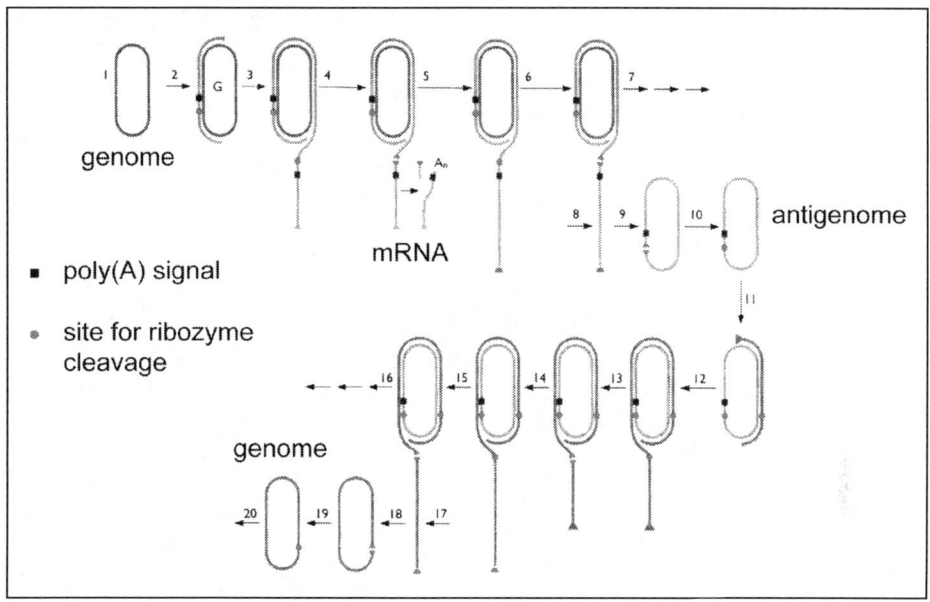

Figure 3. Taylor et al model of HDV replication. See text for explanation. This figure represents further adaptation of a model first presented in 1999[98] and then adapted in 2000.[101] It is shown here with permission.

recombination between viroids, but it must be remembered that viroids are much smaller and have no open reading frame.

Another form of template-switching is intra-molecular rather than inter-molecular. Such switching is an essential part of the life cycle for many viruses, such as retroviruses and hepadnaviruses. It could be reasonably argued that such events are not needed for HDV if the only templates used are circular. That is, rolling-circle models such as those in Figures 3 and 4, would seem to obviate the need for template-switching. As an alternative opinion, it is thus relevant that a recent study in which replication was forced to initiate with HDV RNAs that were not circular but linear, was able to prove that template-switching could occur.[60] Consistent with this interpretation was that for some linear RNAs that were almost exactly unit-length, there arose circular RNA transcripts that had undergone at the discontinuity site, small deletions of HDV sequences and even insertions of nonHDV sequences. In such cases, the template switching is considered to have been imprecise.

Post-Transcriptional Processing

The three main RNAs of HDV, as represented in Figure 1, are all the products of post-transcriptional RNA processing. Both the genome and antigenome contain ribozyme domains. Thus greater than unit-length multimers of genomic and antigenomic RNA can undergo self-cleavage, followed by ligation, to produce unit-length circles. Antigenomic RNAs can undergo three additional forms of RNA-processing. The 5'-end of the mRNA can be capped. The 3'-end of the mRNA is defined by poly(A) processing. And, for some of the antigenomic RNAs, they can be targets for RNA-editing by a host adenosine deaminase.

It is thus obvious that HDV RNAs offer some very interesting cases of post-transcriptional RNA processing. In addition, such processing has to be regulated. And, after the RNAs have been processed to their mature forms, there is the chance for additional processing by host

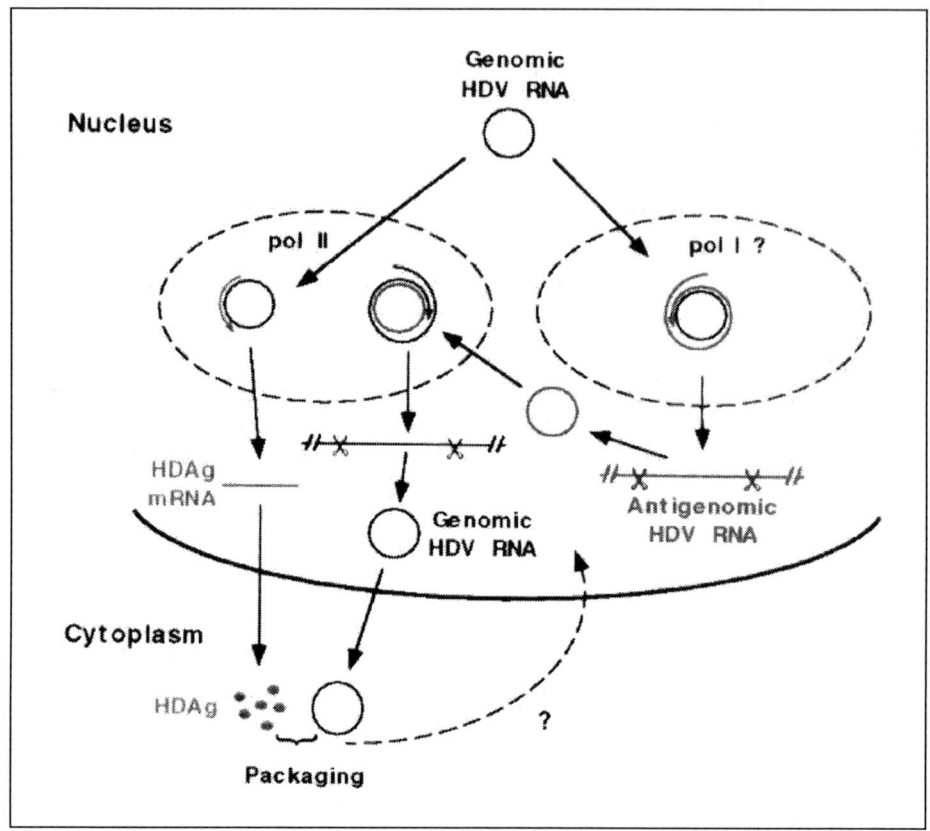

Figure 4. Macnaughton et al model of HDV replication. See text for explanation. This is the model of Macnaughton et al[19] and is shown here with permission.

endonucleases. However, as will be explained, there seems to be no involvement of dicer, a key endonuclease involved in the generation of small interfering RNAs (siRNA).

As explained below and in Figure 2, a good example for discussion of HDV RNA processing is to consider what events can occur on a nascent antigenomic RNA. (Relative to this, discussion of genomic RNA processing has two problems. First, we have no idea of where the transcripts start and second, there is not the complication of poly(A) processing.)

Polyadenylation

For a long time we have known that during replication of HDV RNA there arises a less than full-length RNA.[16] There is indirect evidence that this RNA is 5'-capped[14] and there is direct evidence that it is 3'-polyadenylated.[15] Puzzling features surround this mRNA.

The first concerns its abundance. In most studies the amount of this mRNA relative to the antigenomic RNA is 2% or less. In some experimental situations, the mRNA even needs to be detected by a PCR procedure rather than by poly(A)-selection and Northern analysis. There are two obvious explanations. The processing may be relatively inefficient. Alternatively, the RNA, once processed, may be relatively unstable.

It is plausible that the polyadenylation may be much less than 100% efficient. After all, the transcripts are being made from a RNA rather than a DNA template. How does this polyadenylation compare with the typical polyadenylation of DNA-directed RNA transcripts? In early experiments this question was addressed but at the time, there was a much less clear picture of what is required. Now for example, we would consider it likely that transcription will proceed much further beyond the CA acceptor site before there is (i) pausing, (ii) separation of the polymerase from the template, (iii) endonucleolytic cleavage of the paused RNA, and (iv) specific cleavage at the CA site, (v) finally followed by polyadenylation. On top of this standard picture, we have to superimpose the consequences of there being for HDV, a ribozyme located just 3' of the CA acceptor. (Some of these features are indicated in Fig. 2.) Does this interfere with transcriptional pausing? Does it reduce the efficiency of the polyadenylation? More experiments are needed.

Ribozyme Cleavage

For some time, we have known that both the genomic and antigenomic RNAs contain a domain that will act as a ribozyme.[61,62] These two ribozymes can be reduced to a minimal contiguous sequence of about 85 nt.[63] Many studies have been made to characterize these ribozymes both in vivo and in vitro. The cleavage reaction is a trans-esterification reaction leading to the production of ends with a 5'-OH and a 2', 3'-cyclic monophosphate.[61] The two ribozymes share many sequence features although as enzymes, they have some quite different properties.[64] An atomic structure has been determined for the genomic ribozyme.[65] This has provided the basics for some very detailed studies of the mechanism involved in the cleavage.

Studies have been reported to address the question of whether HDV genome replication can proceed in the absence of one or the other of these two ribozymes.[66,67] The studies so far indicate no accumulation of processed RNAs when either ribozyme is inactivated. In some ways this is puzzling because for many of the plant viroids, the replication does not demand that both strands are processed to unit-length circular species; that is, it can be that just one polarity has such circular RNAs.

Several studies have addressed how sequences outside the ribozyme domains, can affect activity. In vitro, there is no question that adjacent 5' or 3' sequences can interfere with the activity of a ribozyme domain.[62,68] It is interpreted that such extra sequences allow alternative foldings that are inconsistent with a fully active ribozyme.[69] In vivo studies support the interpretation that alternative folding of the ribozyme domain into the predicted rod-like structure interferes with the ribozyme. Furthermore, it has been rationalized that without this inhibition, HDV RNA circles, once formed, would then self-cleave to become linears. The other side of the rod-like structure is thus considered to act as a cis-acting attenuator of the ribozyme. As further support of this model, a short RNA sequence which can 100% base pair with the attenuator can act as a trans-acting anti-attenuator, and allow circles to self-cleave.[69] The anti-attenuator concept has been shown to even work during HDV RNA transcription in vitro by a phage polymerase. Specifically, a DNA oligonucleotide can function as an anti-attenuator to allow >90% in vitro cleavage during transcription (Lazinski, personal communication, with confirmation by Nie and JMT, unpublished observations). A corollary of this is that a small oligonucleotide which binds to the ribozyme, can act as an attenuator during transcription, giving >90% inhibition of self-cleavage.[70] Further experiments are needed to test the effect of such trans-acting oligonucleotides on replication in vivo.

Circularization

The genomic and antigenomic RNAs of HDV are the only examples of animal virus RNAs that are circular in conformation. There are data to support the hypothesis that such circularity provides an advantage in terms of RNA stability. In one controlled study using cell extracts, circles were 300-times more stable than the corresponding linear species.[71]

Of course, the circular conformation has been incorporated into models of HDV genome replication. It is considered that for both the genome and antigenome templated transcription, a double rolling-circle mechanism applies (as in Figs. 3 and 4). Similar models were previously applied to explain the replication of some of the viroid RNAs.[72]

A puzzling issue remains as to how HDV RNAs, after cleavage by their ribozymes, can be circularized. One series of studies led to the speculation that a host factor, probably an RNA ligase, is needed for this circularization.[73] Consistent with this hypothesis, no one has yet been able to achieve in vitro the combination of HDV ribozyme cleavage and circularization, all in the absence of any host protein. (There was a single report of this being achieved,[74] but subsequently it was found that the interpretation of the data was incorrect.) In some ways this is puzzling because for other ribozymes, the combination of protein-free cleavage followed by ligation has been reported.[75,76]

Endonuclease Attack

Even though circularization does enhance the stability of the HDV RNAs, they are nevertheless susceptible to a level of endonucleolytic attack. Some studies have been interpreted as evidence that the presence of δAgs, which are RNA-binding proteins, can enhance the stabilization.[21]

One study reported that for genomic RNA present in the liver of an HDV infected woodchuck, it can be that many of the unit-length RNAs are opened up at a specific site.[18] The mechanism of opening is not due to the HDV ribozyme but is otherwise unexplained.

Recently much attention has been drawn to a host-encoded endonuclease known as dicer.[77] This activity, which is predominantly cytoplasmic, is able to convert double-stranded RNA to fragments of about 21 nt. These fragments are known as small interfering RNAs or siRNA. They are known to participate in a process known as RNA interference, RNAi, in which related mRNA species present in the cytoplasm can be specifically targeted for degradation.[57] A recent study has looked without success for HDV-related siRNA during HDV genome replication.[78] Also, from in vitro experiments with purified dicer, the HDV RNAs were again resistant.[78] A major level of protection seems to be dependent on the long un-branched rod-like folding of the HDV RNAs. This folding is predicted to be 74% rather than 100% of all the nucleotides. These studies show that the resistance can be explained independent of the accessibility of the RNA to dicer, or the presence of δAg or any host protein.

It is possible to produce siRNA in vitro and then transfect these into mammalian cells, leading to targeted degradation. Already this strategy has been used to interfere with the replication of several different animal viruses.[79-82] This strategy was applied using HDV-specific siRNAs, as transfected into cultured cells in which the HDV genome was replicating. It thus was possible to target the HDV mRNAs sequences and to achieve, indirectly, inhibition of the accumulation of HDV genomic and antigenomic RNAs.[83]

ADAR Attack

From the first studies of HDV it was realized that there were two main size classes of δAg that could be resolved by gel electrophoresis.[84] These were given the obvious names of small (δAg-S) and large δAg (δAg-L). From nucleotide sequencing studies it was then noted that for some RNAs the amber codon used for the termination of translation of the 195 amino acid δAg-S, was changed to the codon was for tryptophan, allowing the translation of a larger form of δAg, with an extra 19 amino acids.[85]

Surprisingly, when a chimpanzee was transfected with an infectious clone encoding δAg-S, there appeared during the subsequent HDV replication, an increasing amount of δAg-L.[40] This was soon shown to be due to the accumulation of a nucleotide change at the location corresponding to the middle of the amber termination codon.[86] This is at position 1012 on the sequence of Kuo et al.[13]

Ultimately it was made clear that this change was achieved by the action on the antigenomic RNA of an RNA-editing enzyme, of the class now known as adenosine deaminases acting on RNA or ADAR.[87] These ADAR, in addition to a deaminase activity, have the ability to bind to RNA molecules that are either 100% double-stranded or contain regions that are almost 100%. For HDV there is now clear evidence that the editing site is defined by a patch of rod-like structure surrounding position 1012 on the antigenome.[43] The ADAR converts this adenosine to inosine. After a subsequent round of transcription the UAG amber codon is thus replaced with UGG, encoding tryptophan.

Recent studies have determined the minimum structural element needed for specific ADAR editing.[88] Also, it has been possible to show, at least in one cultured cell line, that the editing enzyme is the gene product of what is known as ADAR-1. Furthermore, there are two main forms of the ADAR-1 protein in mammalian cells, and the HDV editing is via the form that is smaller and predominantly located in the nucleus.[89]

As explained in more detail in Chapter 4, both δAg-S and δAg-L are essential for HDV replication. The small form is essential for the initiation and maintenance of genome replication.[90] The large form is essential for the assembly of HDV genomic RNA into particles with an outer envelope of HBV surface proteins.[91] Thus, the appearance of the large form, as a consequence of ADAR editing, has to be considered as an essential event in the replication cycle.

There is evidence that other sites on the HDV RNAs, both genomic and antigenomic, can undergo ADAR editing.[92,93] However, these other sites are less site-specific and less efficiently used. Additional evidence indicates that the presence of δAg can act as a negative regulator of ADAR editing of HDV RNA.[92] In Chapter 5, ADAR-editing is discussed in greater detail.

RNA Assembly

In natural coinfections of hepatocytes with HBV and HDV, the surface proteins of HBV facilitate the assembly of HDV genomic RNA into particles. As discussed in Chapter 2, this assembly is mediated by δAg-L that has the dual abilities to interact with the HDV RNA and with the HBV envelope proteins. Assembly can also be achieved by transient transfection. Using cells transfected to express the envelope proteins of HBV, there is the assembly into virus-like particles of HDV genomic RNAs created by RNA-directed genome replication.[91,94] Sureau and co-workers first showed that such assembled particles are able to infect cultures of primary hepatocytes.[95] As for HBV, this infectivity depends on the presence of the large form of the HBV surface antigen.[96]

Models of Genome Replication

Figures 3 and 4 show the only two replication schemes that have been published for HDV replication. In this section we will consider some of the complications that can be ascribed to these schemes.

Consider first the model proposed by Macnaughton et al[19] as shown in Figure 4. It incorporates features from an earlier study by Modahl and Lai[39] from which it was proposed that HDV RNA replication and mRNA transcription occur independently and in parallel during the viral life cycle. Two additional features drive this model to be different from the other model. First, the authors try to incorporate their interpretation that while the mRNA and genome are transcribed by pol II, they consider that the antigenomic RNA is transcribed by a different polymerase, possibly pol I. This then leads the authors to indicate that the genomic RNA can serve as template for two polymerases; pol II copies it to produce mRNA then this same template is used by pol I. The authors do not discuss how it might be that the same template can be transcribed by alternative polymerases. The second feature that the model tries to incorporate is the separate finding by these authors[97] of processed unit-length genomic RNAs in the cytoplasm soon after synthesis, which is considered to occur in the nucleus. The processed antigenomic RNAs do not so move to the cytoplasm and this is interpreted in the model as a contributing factor to the observed specificity with which genomic rather than antigenomic RNA is assembled into particles.

The second model, shown in Figure 3, has been passed down following early studies in my lab.[98] This model does not complicate the issue by invoking more than one polymerase or by trying to account for nuclear-cytoplasmic distributions. It focuses only on the transcription and RNA-processing. However, as will now be explained it has at least 3 serious problems.

i. Throughout its 22 steps, it is presumed (for this and the other model) that the only RNAs that can act as templates for transcription are RNA circles. As discussed earlier, this may well not be true. Many experimental studies have shown that linear forms of HDV RNA, genomic and antigenomic, can act as templates to initiate replication. It may well be that the circular forms are more stable and accumulate better, but it is not yet proven that they are the only, or even that they are better templates. It is pertinent to note here that for many of the plant viroids, RNA circles are found for only one polarity; the other polarity is represented only by linear multimers. In other words, linear RNAs can be the only choice of template.

ii. It is presumed that transcription on the genomic RNA template is initiated at a site corresponding to the 5'-end of the mRNA. Even if we accept this, in steps 1-4 it is proposed that first there is transcribed enough antigenomic RNA to allow the RNA-processing to produce the polyadenylated mRNA. Then in steps 5-10, following elongation of what has been called "the continuing transcript", there is generated an RNA that is greater than unit-length, and contains more than one ribozyme. It is considered that ribozyme cleavage followed by ligation leads to the formation of new antigenomic RNA circles. Since this model was first proposed much has been learned about the poly(A)-processing mechanism associated with host DNA-directed pol II transcripts.[99] Apparently the transcription can proceed even 2 kb beyond the poly(A) signals before transcription is terminated and complete assembly of the poly(A) machinery leads to cleavage at the CA acceptor site and a poly(A) polymerase adds the poly(A) tail. In all this, the pol II provides essential scaffolding functions via the multiple tandem repeats in the C-terminal domain of its large protein subunit. Of relevance to HDV, is that the RNA transcripts can reach sizes greater than unit-length before poly(A)-processing (or ribozyme processing) has had a chance to occur. Recent studies of DNA-directed pol II transcripts of HDV antigenomic RNA, in fact, support this interpretation for HDV (Nie, Chang, and JMT, unpublished observations). Furthermore, rather than the concept of the continuing transcript being processed in two different ways, they instead support a new model, namely that that for each antigenomic transcript one of three mutually exclusive alternative outcomes are possible. (a) Some will become poly(A)-processed. (b) Others will be ribozyme-processed, leading to ligation and the formation of unit-length circles. (c) Yet others, and this could even be the major class, will either be not processed at all, or will be processed incompletely or in alternative ways.

iii. The third major problem is that of initiation of transcription. As discussed earlier, apart from our presumption for the 5'-end of the mRNA being an initiation site, we know nothing more about the initiation of transcription. Much more needs to be done to understand initiation sites and efficiencies for HDV genomic and antigenomic RNA templates. Maybe such information will surface from in vitro experiments when we begin to use better templates, such as ribonucleoprotein complexes released from virions.

In summary, these replication schemes contain significant flaws and also fail to show some of the inherent complexity of HDV replication. Nevertheless, they serve a valid purpose if they are able to drive us to specific experimental tests and to refinements, as proven necessary.

Conclusions and Outlook

While the significance of HDV as a threat to human health is decreasing, its interest as an intriguing problem in molecular virology continues to increase. From this and other chapters, it should be clear that we have made serious advancements in our understanding the mechanism of replication of HDV but there remain some important problems yet to be solved. Among these, the two most important may be the remaining confusion about which polymerases are used and our inability to demonstrate the initiation of RNA-directed transcription in vitro.

Acknowledgments

Constructive comments on the manuscript were given by Michael Lai, William Mason, Severin Gudima, Jinhong Chang, and Chi Tarn. This work was supported by grants AI-26522 and CA-06927 from the N.I.H., and by an appropriation from the Commonwealth of Pennsylvania.

References

1. Rizzetto M, Canese MG, Arico J et al. Immunofluorescence detection of a new antigen-antibody system associated to the hepatitis B virus in the liver and in the serum of HBsAg carriers. Gut 1977; 18:997-1003.
2. Gaeta GB, Stroffolini T, Chiaramonte M et al. Chronic hepatitis D: A vanishing disease? An Italian multicenter study. Hepatology 2000; 32:824-827.
3. Casey JL. Hepatitis delta virus: Genetics and pathogenesis. Clin Lab Med 1996; 16:451-464.
4. Gerin JL, Casey JL, Purcell RH. Hepatitis Delta Virus. In: Knipe DM, Howley PM, eds. Fields' Virology. Vol 2. Philadelphia: Lippincott Williams & Wilkins, 2001:3037-3050.
5. Lai MMC. The molecular biology of hepatitis delta virus. Ann Rev Biochem 1995; 64:259-286.
6. Taylor JM. Replication of human hepatitis delta virus: Recent developments. Trends in Microbiol 2003; 11:185-190.
7. Taylor JM. Hepatitis delta virus. Intervirology 1999; 42:173-178.
8. Taylor JM. Replication of human hepatitis delta virus: Influence of studies on subviral plant pathogens. Adv Vir Res 1999; 54:45-60.
9. Monjardino J. Molecular Biology of Human Hepatitis Viruses. London: Imperial College Press, 1998.
10. Modahl LE, Lai MM. Hepatitis delta virus: The molecular basis of laboratory diagnosis. Crit Rev Clin Lab Sci 2000; 37:45-92.
11. Wang K-S, Choo Q-L, Weiner AJ et al. Structure, sequence and expression of the hepatitis delta viral genome. Nature 1986; 323:508-513.
12. Makino S, Chang MF, Shieh CK et al. Molecular cloning and sequencing of a human hepatitis delta (δ) virus RNA. Nature 1987; 329:343-346.
13. Kuo MY-P, Goldberg J, Coates L et al. Molecular cloning of hepatitis delta virus RNA from an infected woodchuck liver: Sequence, structure, and applications. J Virol 1988; 62:1855-1861.
14. Gudima S, Wu S-Y, Chiang C-M et al. Origin of the hepatitis delta virus mRNA. J Virol 2000; 74:7204-7210.
15. Hsieh S-Y, Chao M, Coates L et al. Hepatitis delta virus genome replication: A polyadenylated mRNA for delta antigen. J Virol 1990; 64:3192-3198.
16. Chen P-J, Kalpana G, Goldberg J et al. Structure and replication of the genome of hepatitis δ virus. Proc Natl Acad Sci USA 1986; 83:8774-8778.
17. Gudima S, Dingle K, Wu T-T et al. Characterization of the 5'-ends for polyadenylated RNAs synthesized during the replication of hepatitis delta virus. J Virol 1999; 73:6533-6539.

18. Chang J, Moraleda G, Gudima S et al. Efficient site-specific nonribozyme opening of hepatitis delta virus genomic RNA in infected liver. J Virol 2000; 74:9889-9894.
19. Macnaughton TB, Shi ST, Modahl LE et al. Rolling circle replication of hepatitis delta virus RNA is carried out by two different cellular RNA polymerases. J Virol 2002; 76:3920-3927.
20. Kos A, Dijkema R, Arnberg AC et al. The hepatitis delta (δ) virus possesses a circular RNA. Nature 1986; 323:558-560.
21. Lazinski DW, Taylor JM. Expression of hepatitis delta virus RNA deletions: cs and trans requirements for self-cleavage, ligation, and RNA packaging. J Virol 1994; 68:2879-2888.
22. Loss P, Schmitz M, Steger G et al. Formation of a thermodynamically metastable structure containing hairpin II is critical for infectivity of potato spindle tuber viroid RNA. EMBO J 1991; 10:719-727.
23. Branch AD, Lee SE, Neel OD et al. Prominent polypurine and polypyrimidine tracts in plant viroids and in the RNA of the human hepatitis delta agent. Nucl Acids Res 1993; 21:3529-3535.
24. Branch AD, Benenfeld BJ, Baroudy BM et al. An ultraviolet-sensitive RNA structural element in a viroid-like domain of the hepatitis delta virus. Science 1989; 243:649-652.
25. Circle DA, Lyons AJ, Neel OD et al. Recurring features of local tertiary structure elements in RNA molecules exemplified by hepatitis D virus RNA. RNA 2003; 9:280-286.
26. Branch AD, Benenfeld BJ, Robertson HD. Ultraviolet light-induced crosslinking reveals a unique region of local tertiary structure in potato spindle tuber viroid and HeLa 5S RNA. Proc Natl Acad Sci USA Oct 1985; 82:6590-6594.
27. Robertson HD, Manche L, Mathews MB. Paradoxical interactions between human hepatitis delta agent RNA and the cellular protein kinase PKR. J Virol 1996; 70:5611-5617.
28. Heilman-Miller SL, Woodson SA. Effect of transcription on folding of the Tetrahymena ribozyme. RNA 2003; 9:722-733.
29. Rizzetto M, Canese MG, Gerin JL et al. Transmission of the hepatitis B virus-associated delta antigen to chimpanzees. J Infect Dis 1980; 141:590-602.
30. Ponzetto A, Cote PJ, Popper H et al. Transmission of the hepatitis B virus-associated δ agent to the eastern woodchuck. Proc Natl Acad Sci USA 1984; 81:2208-2212.
31. Netter HJ, Kajino K, Taylor J. Experimental transmission of human hepatitis delta virus to the laboratory mouse. J Virol 1993; 67:3357-3362.
32. Ohashi K, Marion PL, Nakai H et al. Sustained survival of human hepatocytes in mice: A model for in vivo infection with human hepatitis B and hepatitis delta viruses. Nature Medicine 2000; 6:327-331.
33. Taylor J, Mason W, Summers J et al. Replication of human hepatitis delta virus in primary cultures of woodchuck hepatocytes. J Virol 1987; 61:2891-2895.
34. Sureau C, Jacob JR, Eichberg JW et al. Tissue culture system for infection with human hepatitis delta virus. J Virol 1991; 65:3443-3450.
35. Kuo MY-P, Chao M, Taylor J. Initiation of replication of the human hepatitis delta virus genome from cloned DNA: Role of delta antigen. J Virol 1989; 63:1945-1950.
36. Glenn JS, Taylor JM, White JM. In vitro-synthesized hepatitis delta virus RNA initiates genome replication in cultured cells. J Virol 1990; 64:3104-3107.
37. Dingle K, Bichko V, Zuccola H et al. Initiation of hepatitis delta virus genome replication. J Virol 1998; 72:4783-4788.
38. Bichko V, Netter HJ, Taylor J. Introduction of hepatitis delta virus into animal cell lines via cationic liposomes. J Viro 1994; 68:5247-5252.
39. Modahl LE, Lai MMC. Transcription of hepatitis delta antigen mRNA continues throughout hepatitis delta virus (HDV) replication: A new model of HDV RNA transcription and regulation. J Virol 1998; 72:5449-5456.
40. Sureau C, Taylor J, Chao M et al. A cloned DNA copy of hepatitis delta virus is infectious in the chimpanzee. J Virol 1989; 63:4292-4297.
41. Rapicetta M, Ciccaglione AR, D'Urso N et al. Chronic infection in woodchucks infected by a cloned hepatitis delta virus. In: Hadziyannis SJ, Taylor JM, Bonino F, eds. Hepatitis Delta Virus: Molecular Biology, Pathogenesis, and Clinical Aspects. New York: Wiley-Liss, 1993:451.

42. Chang J, Sigal LJ, Lerro A et al. Replication of the human hepatitis delta virus genome is initiated in mouse hepatocytes following intravenous injection of naked DNA or RNA sequences. J Virol 2001; 75:3469-3473.

43. Gudima SO, Chang J, Moraleda G et al. Parameters of human hepatitis delta virus replication: The quantity, quality, and intracellular distribution of viral proteins and RNA. J Virol 2002; 76:3709-3719.

44. Mu JJ, Chen DS, Chen PJ. The conserved serine 177 in the delta antigen of hepatitis delta virus is one putative phosphorylation site and is required for efficient viral RNA replication. J Virol 2001; 75:9087-9095.

45. Darnell J, Lodish H, Baltimore D. Molecular Cell Biology. New York: Scientific American Books Ltd., 1986.

46. Niranjanakumari S, Lasda E, Brazas R et al. Reversible cross-linking combined with immunoprecipitation to study RNA-protein interactions in vivo. Methods 2002; 26:182-190.

47. Fu T-B, Taylor J. The RNAs of hepatitis delta virus are copied by RNA polymerase II in nuclear homogenates. J Virol 1993; 67:6965-6972.

48. Beard MR, Macnaughton TB, Gowans EJ. Identification and characterization of a hepatitis delta virus RNA transcriptional promoter. J Virol 1996; 70:4986-4995.

49. Filipovska J, Konarska MM. Specific HDV RNA-templated transcription by pol II in vitro. RNA 2000; 6:41-54.

50. Yamaguchi Y, Delehouzee S, Handa H. HIV and hepatitis delta virus: Evolution takes different paths to relieve blocks in transcriptional elongation. Microbes Infect 2002; 4:1169-1175.

51. Yamaguchi Y, Filipovska J, Yano K et al. Stimulation of RNA polymerase II elongation by hepatitis delta antigen. Science 2001; 293:124-127.

52. Modahl LE, Macnaughton TB, Zhu N et al. RNA-dependent replication and transcription of hepatitis delta virus RNA involve distinct cellular RNA polymerases. Mol Cell Biol 2000; 20:6030-6039.

53. Schindler I-M, Muhlbach H-P. Involvement of nuclear DNA-dependent RNA polymerases in potato spindle tuber viroid replication: A reevaluation. Plant Sci 1992; 84:221-229.

54. Navarro J-A, Vera A, Flores R. A chloroplastic RNA polymerase resistant to targetitoxin is involved in replication of avocado sunblotch virod. Virology 2000; 268:218-225.

55. Schiebel W, Haas B, Marinkovik S et al. RNA-directed RNA polymerase from tomato leaves. I purification and physical properties. J Biol Chem 1993; 268:11851-11857.

56. Schiebel W, Pelissier T, Riedel L et al. Isolation of an RNA-directed RNA polymerase-specific cDNA from tomato. Plant Cell 1998; 10:2087-2101.

57. Hannon GJ. RNA interference. Nature 2002; 418:244-251.

58. Lai MMC. Genetic recombination in RNA viruses. Curr Topics Microbiol Immunol 1992; 176:21-32.

59. Wu JC, Chiang TY, Shiue WK et al. Recombination of hepatitis D virus RNA sequences and its implications. Mol Biol Evol 1999; 16:1622-1632.

60. Chang J, Taylor J. In vivo RNA-directed transcription, with template switching, by a mammalian RNA polymerase. EMBO J 2002; 21:157-164.

61. Sharmeen L, Kuo MY, Dinter-Gottlieb G et al. The antigenomic RNA of human hepatitis delta virus can undergo self-cleavage. J Virol 1988; 62:2674-2679.

62. Kuo MYP, Sharmeen L, Dinter-Gottlieb G et al. Characterization of self-cleaving RNA sequences on the genome and antigenome of human hepatitis delta virus. J Virol 1988; 62:4439-4444.

63. Perrotta AT, Been MD. A pseudoknot-like structure required for efficient self-cleavage of hepatitis delta virus RNA. Nature 1991; 350:436-436.

64. Shih IH, Been MD. Catalytic strategies of the hepatitis delta virus ribozymes. Ann Rev Biochem 2002; 71:887-917.

65. FerreD'Amare AR, Zhou K, Doudna JA. Crystal structure of a hepatitis delta virus ribozyme. Nature 1998; 395:567-574.

66. Jeng K-S, Daniel A, Lai MMC. A pseudoknot structure is active in vivo and required for hepatitis delta virus RNA replication. J Virol 1996; 70:2403-2410.

67. Macnaughton TB, Wang Y-J, Lai MMC. Replication of hepatitis delta virus RNA: Effect of mutations of the autocatalytic cleavage sites. J Virol 1993; 67:2228-2234.

68. Diegelman-Parente A, Bevilacqua PC. A mechanistic framework for cotranscriptional folding of the HDV genomic ribozyme in the presence of downstream sequence. J Mol Biol 2002; 324:1-16.
69. Lazinski DW, Taylor JM. Regulation of the hepatitis delta virus ribozymes: To cleave or not to cleave. RNA 1995; 1:225-233.
70. Mao Q, Wang S, Li Q. [Inhibition of genomic HDV ribozyme activity by antisense oligodeoxynucleotides]. Zhonghua Yi Xue Za Zhi 1996; 76:30-33.
71. Puttaraju M, Been M. Generation of nuclease resistant circular RNA decoys for HIV-tat and HIV-rev by autocatalytic splicing. Nucl Acids Res 1995; 33:49-51.
72. Branch AD, Robertson HD. A replication cycle for viroids and small infectious RNAs. Science 1984; 223:450-455.
73. Reid CE, Lazinski DW. A host-specific function is required for ligation of a wide variety of ribozyme-processed RNAs. Proc Natl Acad Sci USA 2000; 97:424-429.
74. Diegelman AM, Kool ET. Mimicry of the hepatitis delta virus replication cycle mediated by synthetic circular oligodeoxynucleotides. Chem Biol 1999; 6:569-576.
75. Buzayun J, Gerlach W, Bruening G. Nonenzymatic cleavage and ligation of RNAs complementary to a plant virus satellite RNA. Nature 1986; 323:349-353.
76. Saville B, Collins R. RNA-mediated ligation of self-cleavage products of a Neurospora mitochondrial plasmid transcript. Proc Natl Acad Sci USA 1991; 88:8826-8830.
77. Bernstein E, Caudy AA, Hammond SM et al. Role for a bidentate ribonuclease in the initiation step of RNA interference. Nature 2001; 409:363-366.
78. Chang J, Provost P, Taylor JM. Resistance of human hepatitis delta virus RNAs to dicer activity. J Virol 2003; 77:11910-11917.
79. Bitko V, Barik S. Phenotypic silencing of cytoplasmic genes using sequence-specific double-stranded short interfering RNA and its application in the reverse genetics of wild type negative-strand RNA viruses. BMC Microbiol 2001; 1(1):34.
80. Coburn GA, Cullen BR. Potent and specific inhibition of human immunodeficiency virus type 1 replication by RNA interference. J Virol 2002; 76:9225-9231.
81. Ge Q, McManus MT, Nguyen T et al. RNA interference of influenza virus production by directly targeting mRNA for degradation and indirectly inhibiting all viral RNA transcription. Proc Natl Acad Sci USA 2003; 100:2718-2723.
82. Gitlin L, Karelsky S, Andino R. Short interfering RNA confers intracellular antiviral immunity in human cells. Nature 2002; 418:430-434.
83. Chang J, Taylor JM. Susceptibility of human hepatitis delta virus RNAs to small interfering RNA action. J Virol 2003; 77:9728-9731.
84. Bergmann KF, Gerin JL. Antigens of hepatitis delta virus in the liver and serum of humans and animals. J Infect Dis 1986; 154:702-706.
85. Weiner AJ, Choo Q-L, Wang K-S et al. A single antigenomic open reading frame of the hepatitis delta virus encodes the epitope(s) of both hepatitis delta antigen polypeptides p24 and p27. J Virol 1988; 62:594-599.
86. Luo G, Chao M, Hsieh S-Y et al. A specific base transition occurs on replicating hepatitis delta virus RNA. J Virol 1990; 64:1021-1027.
87. Casey JL, Gerin JL. Hepatitis D virus RNA editing: Specific modification of adenosine in the antigenomic RNA. J Virol 1995; 69:7593-7700.
88. Sato S, Wong SK, Lazinski DW. Hepatitis delta virus minimal substrates competent for editing by ADAR1 and ADAR2. J Virol 2001; 75:8547-8555.
89. Wong SK, Lazinski DW. Replicating hepatitis delta virus RNA is edited in the nucleus by the small form of ADAR1. Proc Natl Acad Sci USA 2002; 99:15118-15123.
90. Chao M, Hsieh S-Y, Taylor J. The antigen of hepatitis delta virus: Examination of in vitro RNA-binding specificity. J Virol 1991; 65:4057-4062.
91. Chang FL, Chen PJ, Tu SJ et al. The large form of hepatitis δ antigen is crucial for the assembly of hepatitis δ virus. Proc Natl Acad Sci USA 1991; 88:8490-8494.
92. Polson AG, Ley III HL, Bass BL et al. Hepatitis delta virus RNA editing is highly specific for the amber/W site and is suppressed by the delta antigen. Mol Cell Biol 1998; 18:1919-1926.
93. Netter HJ, Wu T-T, Bockol M et al. Nucleotide sequence stability of the genome of hepatitis delta virus. J Virol 1995; 69:1687-1692.

94. Ryu W-S, Bayer M, Taylor J. Assembly of hepatitis delta virus particles. J Virol 1992; 66:2310-2315.
95. Sureau C, Moriarty AM, Thornton GB et al. Production of infectious hepatitis delta virus in vitro and neutralization with antibodies directed against hepatitis B virus preS antigens. J Virol 1992; 66:1241-1245.
96. Sureau C, Guerra B, Lanford RE. Role of the large hepatitis B virus envelope protein in infectivity of the hepatitis delta virion. J Virol 1993; 67:366-372.
97. Macnaughton TB, Lai MM. Genomic but not antigenomic hepatitis delta virus RNA is preferentially exported from the nucleus immediately after synthesis and processing. J Virol 2002; 76:3928-3935.
98. Taylor JM. Human hepatitis delta virus: Structure and replication of the genome. Curr Top Microbiol Immunol 1999; 239:108-122.
99. Dye MJ, Proudfoot NJ. Multiple transcript cleavage precedes polymerase release in termination by RNA polymerase II. Cell 2001; 105:669-681.
100. Polo JM, Lim B, Govindarajan S et al. Replication of hepatitis delta virus RNA in mice after intramuscular injection of plasmid DNA. J Virol 1995; 69:5203-5207.
101. Flint SJ, Enquist LW, Krug RM et al. Principles of Virology. Washington: ASM Press, 2000.

CHAPTER 4

Hepatitis Delta Antigen:

Biochemical Properties and Functional Roles in HDV Replication

Michael M.C. Lai*

H epatitis delta antigen (HDAg) was first detected in the nucleus of the hepatocytes of some patients infected with hepatitis B virus (HBV).[68] The presence of HDAg was frequently associated with severe hepatitis. This antigen was initially thought to be a previously unrecognized HBV-encoded antigen, but later was found to be associated with a novel virus, hepatitis delta virus (HDV).[69] HDAg is an internal component of the HDV virion particles, and, together with the viral RNA genome, forms viral nucleocapsid.[2] There are approximately 70 HDAg molecules per RNA molecule in each virion particle,[72] but the precise structure of the nucleocapsid has not been determined. HDAg in virus preparations from most of the patients usually consists of two distinct forms of different size (27 and 24 kDa, termed the large and the small HDAg, L- and S-HDAg, respectively). The nucleocapsid is released from the virus particle after the latter is treated with nonionic detergents.[72] Besides this structural role, HDAg also plays a very critical role in the HDV life cycle by participating in various steps of viral replication, including viral RNA synthesis (by S-HDAg), virus assembly (by L-HDAg) and others. HDAg is the only known functional protein encoded by HDV RNA. It is encoded by the antigenomic-sense strand of HDV RNA, but is translated from a 0.8-kb mRNA, which is transcribed from the viral genomic RNA.

Structural and Functional Domains of HDAg

The S-HDAg (195 amino acids) and L-HDAg (214 amino acids) are identical except that L-HDAg has 19 additional amino acids at the C-terminus. The N-terminal two-thirds of the protein are highly basic, while the C-terminal one-third is relatively uncharged.[9] At least four structural and functional domains have been identified in HDAg (Fig. 1).

RNA-Binding Domains

There are several RNA-binding domains within HDAg. The first identified one is mapped in the middle one-third of the protein.[46] This domain consists of two stretches of arginine-rich motif (ARM),[45] which had been identified in several other viral RNA-binding proteins, such as *rev* and *tat* of human immunodeficiency virus.[38] Both of the ARM sequences and a spacer

*Michael M.C. Lai—Department of Molecular Microbiology and Immunology, University of Southern California, Keck School of Medicine, Los Angeles, California 90033, U.S.A.; and Institute of Molecular Biology, Academia Sinica, Taipei, Taiwan 115. Email: michlai@usc.edu

Hepatitis Delta Virus, edited by Hiroshi Handa and Yuki Yamaguchi.
©2006 Landes Bioscience and Springer Science+Business Media.

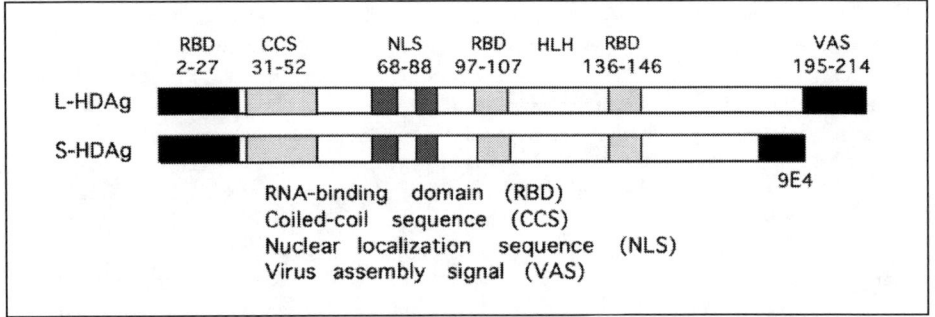

Figure 1. Schematic diagram of the functional domains of the large and small hepatitis delta antigen.

sequence of appropriate length, which contains a helix-loop-helix (HLH) motif,[10] are required for RNA binding.[45] Another potential RNA-binding motif is found in the N-terminal region of the protein (aa 2-27).[65] The RNA-binding activity is the basis of the various biological functions associated with HDAg, such as RNA replication and RNA transport. Studies on the HDAg-mediated HDV RNA transport indicated that any one of these three RNA-binding motifs is sufficient for mediating HDV RNA-HDAg binding,[19] although the requirement for RNA binding in vitro appears to be more stringent.

HDAg is a general RNA-binding protein,[12,34,46] with a particularly high binding affinity for HDV RNA; however, the binding specificity is not absolute. HDAg binds to both genomic and antigenomic strands equally efficiently. In the virion particles, HDAg is bound to HDV RNA,[46] which is exclusively of genomic strand. Thus, there are other factors which contribute to the specificity of RNA packaging into the virus particles. The HDAg-RNA interaction plays a role not only in the structural organization of virion particles, but also in HDV RNA replication (see below).

Nuclear Localization Signal (NLS)

The main NLS of HDAg is located in the N-terminal one-third (aa 68-88) of the protein.[88] This domain contains two stretches of basic amino acids, both of which are required for targeting HDAg to the nuclei. HDAg carries the incoming HDV RNA to the nucleus, where RNA replication occurs.[19] HDAg utilizes the classical nuclear-importing machinery of karyopherin α2β pathway.[19] It has been shown that the HDV RNA-HDAg complex can shuttle in and out of the nucleus,[77] suggesting that HDAg may also have a nucleus-exporting function. However, so far, the nucleus-import and -export domains within HDAg can not be separated. Significantly, a nucleus-exporting signal has been identified in the C-terminal 19 amino acids of L-HDAg, which allows L-HDAg to be located in the cytoplasm to interact with HBsAg.[42] This domain by itself can serve as a nuclear export signal in Xenopus oocytes. This property is consistent with the functional role of L-HDAg in virus assembly, which likely takes place in the cytoplasm. Obviously, there must be a second nucleus-exporting signal which is present in both S- and L-HDAg to account for the export of HDV RNA (particularly the genomic strand) immediately after synthesis.[54] Since HDV RNA likely does not have an intrinsic nuclear localization signal, the final localization of HDV RNA is likely determined by the properties of HDAg, which complexes with HDV RNA. The nucleus-export of L-HDAg and HDV RNA utilizes a Crm1 (chromosome region maintenance 1)-independent pathway.[42] Deletion of the Arg-rich motif also leads to the accumulation of HDV RNA in the cytoplasm, indicating that the nuclear localization of HDV RNA requires the interaction between HDV RNA and HDAg.[19]

Coiled-Coil Sequence

This motif is present in the N-terminal one-third (aa 31-52) of the protein.[15,88] This domain is responsible for oligomerization of the protein. The large and small HDAg can form either homo- or heterodimers of multiple size in vitro.[70,88] The HDAg prepared from the HDV-replicating cells or from the recombinant HDAg preparations usually formed a large complex.[83,73] Interestingly, a peptide (aa 12 to aa 60) corresponding to this coiled-coil domain formed an antiparallel dimer.[93] The oligomerization of HDAg is required for the various functions of both the large and small HDAg, such as the trans-acting function of S-HDAg in HDV RNA replication and the suppression of HDV replication by L-HDAg[88] (see below). Thus, the formation of uniform homodimers appears to be important for the replication of HDV RNA, as mixtures of L- and S-HDAg suppressed viral RNA replication. The helix-loop-helix (HLH) sequence in the middle of HDAg[10] may also contribute to the complex formation of HDAg.

The C-Terminal 19-Amino Acid Extension of L-HDAg

This domain is present only in L-HDAg, and represents a stretch of sequence important for virus assembly (virus assembly signal, VAS). This sequence is required for the interaction of L-HDAg with HBsAg,[21,33] which is a critical step in HDV particle assembly. The 19 amino acids are necessary and sufficient as the signal for virion assembly.[43] Curiously, this 19-amino-acid sequence is extremely divergent among the three HDV genotypes.[5] This region is prenylated at cystine-211,[24,34,44,64] consistent with the known prenylation motif (CXXX).

At the very C-terminus of S-HDAg (aa 146-195) and the corresponding region of L-HDAg is a stretch of sequence rich in proline and glycine and is relatively hydrophobic. So far, no function has been assigned to this region, but it is relatively conserved. Site-specific mutagenesis studies indicate that this domain is also indispensible for HDV RNA replication.[40] The conformation of this domain can be modified by the presence of the last 19 amino acids in L-HDAg. For example, a C-terminal epitope (9E4) that is present in S-HDAg is not detected in L-HDAg, even though the primary sequence is present.[34a]

Post-Translational Modifications

HDAg is modified by phosphorylation, acetylation, prenylation and methylation. These post-translational modifications play very significant roles in determining the structure of viral particles and in regulating various steps of HDV life cycle.

Phosphorylation

Both S- and L-HDAg are phosphorylated at multiple sites, mostly at serine and threonine residues.[9,63] Serine-177 is the predominant phosphorylation site in S-HDAg in vivo. Phosphorylation has been reported to be carried out by casein kinase II, protein kinase C (PKC) or double-stranded RNA-activated kinase PKR and is required for certain steps of HDV RNA replication.[14,92] However, the relative significance of each kinase in the phosphorylation of HDAg is not clear. The PKC-specific inhibitor H7 has been shown to decrease the phosphorylation of S-HDAg and suppress HDV RNA replication, indicating that phosphorylation is a positive regulator of HDV RNA replication.[92] On the other hand, in cells expressing a dominant-negative mutant of PKR, the level of phosphorylation of S-HDAg was suppressed but the level of HDV RNA replication was enhanced, suggesting that the phosphorylation of HDAg by PKR may serve as a negative regulator of HDV replication.[14,61] However, the S177A mutant has a significantly lower level of HDV RNA replication; specifically, this mutation interferes with the replication of the HDV antigenomic RNA to genomic RNA, but not the replication of genomic RNA to antigenomic RNA, suggesting that the synthesis of the two strands of HDV RNA have different requirements for the phosphorylation of S-HDAg. S177A mutation also affects the efficiency of RNA editing. These contradicting data indicate that the role of HDAg phosphorylation in HDV replication is very complex. The significance of other phosphorylation sites, such as S2 and S123,[92] is not clear. Interestingly, the phosphorylation status of S-HDAg in the HDV-replicating cells is different from that in the released HDV particles;[61] for example, S-HDAg in the virion is not phosphorylated, whereas multiple phosphorylated species are detected in the cells, indicating that only the unphosphorylated S-HDAg species can be incorporated into virus particles. By contrast, L-HDAg in the virion is phosphorylated. Therefore, phosphorylation and dephosphorylation of HDAg are important not only for HDV RNA replication, but also for virus assembly.

Prenylation

L-HDAg is prenylated at the C-211 residue. The nature of prenylate has been identified as farnesylate.[64] Farnesylation of L-HDAg has a number of functional consequences. First, prenylation of proteins is generally thought to facilitate membrane association of proteins, a possibility that would facilitate HDV virion assembly. However, prenylation of L-HDAg has been shown to enable the direct interaction between L-HDAg and HBsAg,[21,33] which is a critical step in HDV particle assembly. Indeed, prenylation at C-211 is necessary for the formation of HDV particles.[24] Prenylation inhibitors inhibit the production of infectious virus particles.[3] Second, prenylation alters the conformation of HDAg, resulting in the masking of a unique epitope (9E4),[32] which is mapped to the exact C-terminal end of S-HDAg. This epitope can not be detected in L-HDAg, even though the primary sequence of this epitope is present in L-HDAg. When the prenylate is removed, this epitope is exposed in L-HDAg. Interestingly, this epitope in L-HDAg also could not be recognized by the humoral arm of the immune system when it was inoculated into mice.[30] Third, prenylation enhances the inhibitory activity of L-HDAg on HDV RNA replication, when L-HDAg was coexpressed with S-HDAg early in the replication cycle.[32] This result was likely due to the prenylate-induced conformational change of L-HDAg, causing L-HDAg to become even more diverged from S-HDAg in its structure; consequently the replication complex formed from such an L-HDAg is not functional.

Acetylation

Recent studies have shown that S- and L-HDAg are also acetylated.[62] Acetylation occurs at multiple sites, among which Lys-72 appears to be the primary site. It is carried out by cellular acetyltransferase p300. Acetylation of Lys-72 is necessary for nuclear transport of HDAg; substitution of Lys-72 to Arg caused the redistribution of HDAg from the nucleus to the cytoplasm. Intriguingly, this amino acid substitution reduced viral RNA accumulation and resulted in earlier appearance of L-HDAg, suggesting that the posttranslational modification of Lys-72 affects the regulation of RNA editing. This modification appears to be necessary for nucleocytoplasmic shuttling of HDAg. It is not clear whether the reduced RNA replication is the direct effect of HDAg acetylation on HDV RNA replication or the indirect effect of HDAg redistribution in the cells.

Methylation

Another type of modification of HDAg was recently discovered, namely, methylation.[45a] Both arginine and lysine methylation have been detected. The arginine methylation occurs on Arg-13, which is located within the N-terminal RNA-binding motif. This modification is required for HDV RNA replication, specifically the replication of the antigenomic strand to genomic strand. The R13A mutant is defective in HDV RNA replication. This finding explains the previous observation that the recombinant HDAg derived from E. coli could not initiate RNA replication from the antigenomic HDV RNA using the RNA transfection method.[73] Once the recombinant S-HDAg was methylated in vitro by arginine

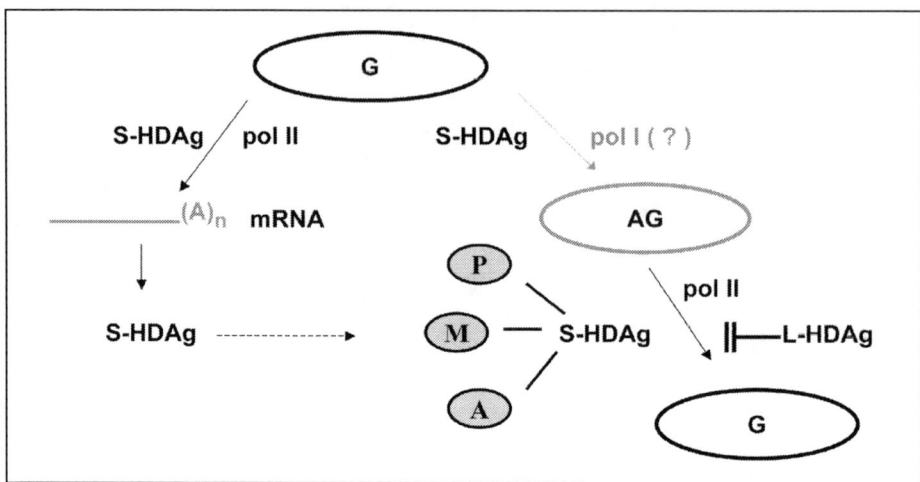

Figure 2. The roles of HDAg in the replication and transcription of HDV RNA. The various forms of HDAg and the possible polymerases involved in the different steps of HDV replication are indicated. The genomic strand is in black and the antigenomic strand is in a lighter color. P: phosphorylation, M: methylation, A: acetylation.

methyltransferase, it became capable of supporting the replication of the HDV antigenomic RNA.[45a] Curiously, the R13A mutation also affected the nucleocytoplasmic transport of HDAg. The R13A mutant is partially localized in the cytoplasm and does not form the speckle pattern characteristic of the wild-type HDAg in the cells. Thus, the nucleocytoplasmic transport of HDAg requires both acetylation and methylation of the protein.

Subcellular Localization of HDAg

In virus-infected cells, HDAg appears to be localized exclusively in the nuclei, based on immunohistochemical staining.[68] When HDAg (S- or L-HDAg) alone was expressed in various cell types, it could be seen variably in the nucleoli, nucleoplasm or diffusely throughout the nuclei, depending on individual cells.[88] Curiously, HDAg in the nucleoli colocalizes with protein kinase PKR.[14] In the cell culture undergoing HDV RNA replication, HDAg is very often present in speckles in the nuclei. Whether these speckles represent the site of active RNA replication is not yet clear. Recent biochemical fractionation and metabolic labeling studies have further shown that a substantial portion of HDAg is present in the cytoplasm (Li and Lai et al, unpublished). Furthermore, HDV ribonucleocapsid can be shuttled between the cytoplasm and the nucleus;[77] this property likely reflects the ability of HDAg to shuttle in and out of the nucleus. In addition to its role in virion assembly, HDAg may perform other functions in the cytoplasm. Particularly, whether S-HDAg participates in certain unrecognized steps of HDV replication in the cytoplasm is an open question.

Functions of HDAg in HDV Replication

Both S- and L-HDAg play very significant roles in the HDV replication cycle. Some of the functions appear to be shared by both proteins, but each HDAg form also possesses distinct functions (Fig. 2).

Trans-Acting Function for HDV RNA Replication (S-HDAg)

S-HDAg is absolutely required for HDV RNA replication in vivo;[37] it acts in trans and complements defects of the HDV RNA mutants that do not synthesize a functional HDAg.[37] It is required for the initiation of the HDV replication cycle. When HDV RNA is transfected into cells, it can not undergo RNA replication unless an mRNA encoding the S-HDAg or a recombinant HDAg protein is cotransfected. Once HDV RNA synthesis is initiated, the subsequent RNA synthesis is sustained by the additional HDAg made from the subgenomic HDV mRNA transcribed from the HDV genomic RNA. An alternative experimental approach is the transfection of HDV RNA or cDNA into cells that express S-HDAg. In still another method, the multimeric HCV cDNA is transfected into cells; HDAg is made from the RNA transcript of the transfected HDV cDNA. In all of these experimental approaches, the production of HDAg is absolutely necessary for continuous HDV RNA replication. The ability of S-HDAg to promote HDV RNA replication appears to be genotype-specific, as S-HDAg of genotype I cannot support genotype III RNA replication and vice versa.[6] This finding suggests that HDAg may recognize certain HDV RNA sequence in order to initiate RNA synthesis.

The mechanism by which HDAg is involved in HDV RNA replication is not clear. There are several possible scenarios. First of all, HDAg may serve as a transcription factor. HDAg is a nuclear protein and contains coiled-coil and HLH domains and a long stretch of proline- and glycine-rich sequence (in the C-terminal region), characteristic of many transcription factors for DNA-dependent transcription factors. Furthermore, it is acetylated and methylated, very similar to the DNA-dependent transcription factors. HDAg also has been shown to interact with RNA polymerase II.[91] However, so far, the role of HDAg in the initiation of HDV RNA replication has not been demonstrated in vitro. In several reported in vitro studies for HDV RNA replication, HDAg was not required and its addition did not affect RNA synthesis.[23,52] The closest evidence to demonstrate such a function so far came from an in vitro pol II-mediated, HDV RNA-templated transcription elongation assay.[91] This study demonstrated that HDAg can facilitate elongation of HDV RNA synthesis. Curiously, HDAg has similar activities for cellular DNA-templated, pol II-mediated transcription.[91] This function is very similar to that of the *tat* protein of HIV.[90] Correspondingly, this study shows that HDAg shares some limited sequence similarity with an elongation factor NELF-A (the subunit A of negative elongation factor). HDAg also interacts with a nucleolar protein B23[29] and several other nuclear transcription factors (Y. Wu-Lee, personal communication); over-expression of B23 enhances HDV RNA synthesis.[29] HDAg is also colocalized with nucleolin, a nucleolar protein.[41] These findings are consistent with the proposal that HDV RNA replication is mediated by both pol II (for genomic RNA and mRNA transcription) and pol I (for antigenomic RNA replication), based on the differential sensitivity of the synthesis of these two strands of RNA to α-amanitin.[57,60]

The second possibility is that HDAg binds to HDV RNA and alters the conformation of HDV RNA, so that HDV RNA may be recognized by cellular DNA-dependent RNA polymerases. It has been shown that HDV cDNA itself can serve directly as a promoter and be transcribed by cellular polymerases in mammalian cells.[50,76] In fact, one of the promoter elements in HDV cDNA is mapped to one end of the proposed rod-like structure of HDV RNA, corresponding to the promoter for RNA-dependent transcription.[1] Thus, HDAg may recognize the same structural elements on both DNA and RNA. HDAg may change the conformation of HDV RNA so that the RNA mimics the conformation of the native DNA promoter sequence.

The third possibility is that HDAg may not directly participate in HDV RNA replication. Instead, S-HDAg may merely serve as a carrier to transport HDV RNA into the nucleus,[52] where HDV RNA replication takes place.

Trans-Dominant Suppression of RNA Synthesis (L-HDAg)

L-HDAg has been demonstrated to suppress HDV RNA replication when it was coexpressed with S-HDAg and HDV RNA.[30] Also, L-HDAg, by itself, can not support HDV RNA replication. Thus, L-HDAg was thought to be critical for modulating the amount of HDV RNA in viral life cycle.[13] The ability of L-HDAg to suppress HDV RNA requires the coiled-coil sequence.[88] Therefore, the mechanism of inhibition by L-HDAg can be explained by the possibility that L-HDAg forms heterogeneous complexes with S-HDAg, preventing the formation of the homogeneous complex of S-HDAg. However, recent studies have suggested that this observed inhibitory activity of L-HDAg is an artefact of the experimental approach. The inhibitory activity was noted only when L-HDAg was over-expressed artificially at the very early stage of the HDV replication cycle;[55] even under this condition, L-HDAg interferes only with the synthesis of genomic RNA from the antigenomic RNA template, but not vice versa.[58] Furthermore, it has been shown that the presence or absence of L-HDAg does not affect the final level of HDV RNA in the cells.[55] Nevertheless, it is still possible that this inhibitory effect has some biological roles in HDV replication.

Virus Assembly (L-HDAg)

Another important function of L-HDAg is involved in the assembly of HDV virion particles. The presence of both L-HDAg and the hepatitis B virus envelope protein HBsAg triggers the formation of the virus particles.[7,71,75] A virus-like particle can be formed even in the absence of HDV RNA. The C-terminal 19-amino acid sequence of L-HDAg is necessary and sufficient as the virus-assembly signal.[32,43] The prenylation at C211 is required for this process; prenylation facilitates the interaction between L-HDAg and HBsAg,[33] which is the first step of virus assembly. In contrast, S-HDAg can not trigger HDV virion assembly and its presence is not required for virion assembly. However, S-HDAg can be packaged into virus particles,[15] probably because of its interaction with L-HDAg.[88] Also, the presence of S-HDAg enhances the efficiency of RNA packaging into HDV virion.[82] It has been shown that the L- HDAg by itself can bind to HBsAg in vitro.[32] This interaction requires prenylation and specific amino acid sequences within the C-terminal 19 amino acids of L-HDAg.[24,44] Among the three species of HBsAg, the small HBsAg alone, but not the other forms of HBsAg, is sufficient for HDV particle assembly,[71,75] but the presence of the large HBsAg (preS1) in the assembled particles is required for the infectivity of the virus particles in primary primate hepatocyte cultures.[75]

The relative ratio of L- and S-HDAg in HDV virus particles obtained from infected serum varies from 0.5 to 10 or higher, suggesting that there is no strict requirement for a fixed ratio between the large and small forms of HDAg in virus particles. Recent studies have shown that HDAg can shuttle in and out of the nuclei.[77] It has also been shown that an HDAg mutant with a deletion in the nuclear localization signal could still be packaged into virus particles,[11] suggesting that the L-HDAg-HBsAg interaction takes place in the cytoplasm. RNA packaging into the virion particles occurs as a result of the interaction between HDAg and HDV RNA. Only the genomic-sense RNA is packaged into virion although the antigenomic-sense RNA binds equally well to HDAg in vitro.[12,46] The basis for the selectivity of RNA packaging into virions is not yet clear. In the assembled virus particles, HDAg has been shown to complex with HDV RNA.[46]

RNA Chaperone and Enhancement of Ribozyme Activity

HDAg has been shown to have an RNA chaperone activity,[31] which facilitates the unfolding and refolding of RNA in general. In the case of HDV RNA, HDAg promotes the formation of the ribozyme-active RNA conformation, thereby enhancing its ribozyme activity. Indeed, HDAg has been shown to enhance the turnover rate of ribozyme RNA cleavage in vitro[31] and enhance the extent of ribozyme cleavage in the cells.[36] This effect is not specific for HDV RNA. The RNA chaperone activity is mapped to the N-terminal one-third, specifically within aa 24-75, of HDAg; therefore, in theory, it is associated with both S- and L-HDAg. It requires the RNA-binding properties of HDAg. Curiously, the main RNA-binding activity of HDAg is mapped to the middle region of the protein, which does not have the RNA chaperone activity. It also does not overlap with the N-terminal RNA-binding domain of HDAg. Instead, the RNA-chaperone domain overlaps the coiled-coil domain. In vitro studies did show that the RNA-chaperone domain contained nonspecific RNA-binding properties.[80] The enhancement of the ribozyme activity will likely enhance HDV RNA replication. This is probably another mechanism by which HDAg facilitates HDV RNA replication.

Other Activities

HDAg has also been reported to have other activities, most notably, suppression of polyadenylation of mRNA.[28] This activity was demonstrated by using an artificial DNA plasmid containing an SV40 mRNA initiation and termination signal in vitro. This observation prompted the proposal of an HDV RNA replication model, in which the 0.8-kb mRNA transcription is terminated by HDAg, thus allowing the replication of the full-length HDV RNA.[79] According to this model, the transcription of the 0.8-kb mRNA is the first step of RNA replication; after HDAg is synthesized, the mRNA transcription is inhibited by HDAg, allowing replication to proceed through the polyadenylation site, thus generating the full-length RNA product. However, recent studies have shown that the transcription of mRNA and replication of the full-length RNA are independent.[59] Thus, this long-held hypothesis likely represents artefacts of the experimental systems.

HDAg has been shown to stabilize HDV RNA, probably also because of its RNA-binding properties.[39] The biological significance of this property is not yet clear. In addition, HDAg has been shown to inhibit the extent of RNA editing in a genotype-specific manner.[17]

Regulation of the Synthesis of S- and L-HDAg

HDAg is made from the 0.8-kb antigenomic-sense mRNA.[16,27] This RNA is the only HDV RNA species capable of synthesizing HDAg.[47] Even though the full-length antigenomic RNA contains the entire coding sequence for HDAg, it can not be used for making HDAg, as it is not associated with polysomes.[47] The initial hypothesis for the regulation of the synthesis of this mRNA species proposed that the transcription of this mRNA occurs only at the beginning of the HDV RNA replication cycle, when RNA synthesis is terminated by the polyadenylation signal.[27] As HDAg is synthesized, HDAg exerts a feedback inhibitory effect on further mRNA transcription by suppressing the poly(A) addition. This hypothesis is not adequate to explain how new mRNA species can be made at the later stage of HDV replication cycle, particularly after HDV RNA has been edited and new mRNA has to be made to translate L-HDAg. Recent studies have indicated that the transcription of this mRNA and replication of the full-length HDV RNA are independent and that the 0.8-kb mRNA is continuously transcribed throughout the replication cycle.[59] Thus, HDAg can be made throughout the HDV replication cycle, and the production of the different HDAg species (S- and L-HDAg) can be regulated at the transcription level.

Two different mRNA species are used to synthesize S- and L-HDAg species respectively. The origin of these two different mRNA species is the result of a specific RNA editing event during HDV RNA replication. During HDV RNA replication, a specific nucleotide conversion (U→C) occurs at nucleotide 1015 of the genomic sense RNA, changing the termination codon for small HDAg (on the antigenomic strand) to a tryptophan codon and extending the ORF for an additional 19 amino acids.[49] This mutation results in the synthesis of the L-HDAg.[7,71] This RNA editing appears to be carried out by a double-stranded RNA-adenosine deaminase (ADAR).[67] More recent studies have further identified ADAR I, but not ADAR II, as the enzyme responsible for HDV RNA editing.[35,86] The efficiency of RNA editing in vitro appears to be very high; however, the extent of RNA editing in the cells is regulated, so that L-HDAg is not over-produced. It was previously suggested that L-HDAg itself can suppress RNA synthesis, thus limiting the amount of RNA synthesized at the later stages of viral replication cycle.[13] However, recent studies suggest that the feedback inhibition is probably a result of the enhanced deleterious mutations in the genome triggered by the edited RNA sequence[56] or mediated by L-HDAg per se.[17] These mechanisms can explain how the L-HDAg-encoding mRNA is specifically suppressed. In every delta hepatitis patient examined, both of the RNA species containing a large and a small HDAg ORF are present.[87]

The Roles of HDAg in HDV RNA Synthesis

As discussed above, HDAg is necessary for HDV RNA synthesis. Although the mechanism of its participation in HDV RNA synthesis is still not clear, it is known that the synthesis of genomic and antigenomic RNA strands have differential requirements for HDAg. For example, antigenomic RNA can not initiate RNA replication when it is transfected together with the recombinant HDAg derived from E. coli, whereas genomic RNA can.[73] Phosphorylation (at serine-177) and methylation (at arginine-13) of HDAg are required for the replication of the antigenomic but not the genomic RNA strand.[45a,61] Genomic, but not antigenomic, RNA synthesis is inhibited by L-HDAg, when the latter is expressed early in viral replication.[58] These findings are also consistent with the studies showing differential sensitivity of the genomic and antigenomic RNA synthesis to α-amanitin, i.e., the genomic RNA synthesis (from the antigenomic RNA template) is sensitive to the low concentration of α-amanitin, whereas antigenomic RNA synthesis (from the genomic RNA template) is resistant (up to 100 ug/ml of α-amanitin).[57,60] The mRNA transcription is also sensitive to the low concentration of α-amanitin. Furthermore, Pol II can elongate RNA-dependent RNA synthesis in vitro using HDV RNA as template, although the RNA product is different from the natural RNA species.[22] These findings suggest that polymerase II and an additional polymerase, likely pol I, are both involved in HDV RNA replication. HDAg has been shown to complex with pol II.[91] These studies suggest that HDAg is directly involved in HDV RNA replication, rather than affects the general transcription machinery indirectly. Finally, HDAg also has been reported to complex with a nucleolar protein B23, and HDV RNA synthesis is enhanced when B23 is over-expressed,[29] suggesting that HDAg may be located in the nucleolus at certain stages of the viral life cycle and that this localization is associated with HDV RNA replication.

The question remains as to how RNA polymerase II or other cellular polymerases can utilize an RNA template rather than their normal DNA template. Possibly, HDAg may complex with cellular transcription factors and then interact directly or indirectly with pol II, thereby altering its template specificity. Since HDAg binds to HDV RNA,[9,46] a transcription complex consisting of HDAg, pol II and cellular transcription factors could conceivably complex with HDV RNA and carry out RNA-dependent RNA replication. However, HDV RNA synthesis did occur in the absence of HDAg in cell-free lysates, and the addition of HDAg did not have significant effects on HDV RNA synthesis in vitro.[22,23,52] A more recent study showed that HDAg can promote the elongation, but not initiation, of HDV RNA-dependent, pol II-mediated RNA synthesis in vitro.[91] It is noteworthy that the rod-shaped structure of HDV RNA resembles double-stranded DNA; therefore, it is conceivable that cellular RNA polymerases and transcription factors may recognize double-stranded RNA. Indeed, the double-stranded DNA counterpart of the region encompassing the replication origin for the antigenomic-strand HDV RNA has a promoter activity for transcription.[50,76] How the post-translational modifications of HDAg can affect HDV RNA synthesis is a very interesting question. These questions remain the most critical issues in the study of HDV replication cycle.

The Cytotoxic Effects of HDAg

HDAg, particularly the S-HDAg, has been shown to be cytotoxic when it was expressed at high levels in the cells.[20,51,53,81] However, no significant pathology has been noted in the HDAg-expressing transgenic mice[26,66] and HDAg-expressing cell lines, or associated with HDV infection of primary hepatocyte cultures.[18,74,78] In contrast, both the small and large HDAg have been shown to affect cellular pol II-mediated transcription in vitro and in vivo; in one study, both S- and L-HDAg inhibited pol II-mediated transcription in vitro and in vivo.[48] However, in another study, L-HDAg, but not S-HDAg, stimulated pol II-mediated transcription.[84] L-HDAg has been shown to activate serum response factor-dependent transcription, but not other signal transduction pathways. In contrast, S-HDAg did not have such an effect.[25] The biological significance of these findings is not known. In any case, these effects will likely cause cytotoxicity.

HDV replication has also been shown to induce apoptosis in avian cells.[8] However, in a different study, no apoptosis or cell cycle arrest of HDV-replicating cells were noted.[81] Nevertheless, the cells harboring HDV RNA were gradually lost, suggesting a growth disadvantage associated with HDV replication. The precise effects of HDV replication on the host cells will require further studies.

Perspectives

HDAg plays a very important role in the replication cycle of HDV. It serves as the structural component of the virion and also plays various roles in different steps of the viral replication processes, including RNA replication and virus assembly. The most intriguing aspect of the HDAg biology is its role in converting the cellular DNA-dependent RNA polymerases to an enzyme capable of replicating RNA. This capability opens up a possibility that certain cellular RNA may be replicated through an RNA-dependent process via a protein very similar to HDAg. Indeed, several cellular proteins bear limited sequence similarity with HDAg, including the negative elongation factor (NELF-A) and another uncharacterized protein DIPA (Delta antigen-interacting protein).[4] These proteins may have potential to perform similar functions to that of HDAg. The understanding of these processes will reveal novel features of HDV replication and the molecular biology of mammalian cells.

References

1. Beard MR, Macnaughton TB, Gowans EJ. Identification and characterization of a hepatitis delta virus RNA transcriptional promoter. J Virol 1996; 70:4986-4995.
2. Bonino F, Hoyer B, Shih J et al. Delta hepatitis agent: Structural and antigenic properties of the delta-associated particles. Infect Immunity 1984; 43:1000-1005.
3. Bordier BB, Marion PL, Ohashi K et al. A prenylation inhibitor prevents production of infectious hepatitis delta virus particles. J Virol 2002; 76:10465-72.
4. Brazas R, Ganem D. A cellular homolog of hepatitis delta antigen: Implications for viral replication and evolution. Science 1996; 274:90-94.
5. Casey JL, Brown TL, Colan EJ et al. A genotype of hepatitis D virus that occurs in northern South America. Proc Natl Acad Sci USA 1993; 90:9016-9020.

6. Casey JL, Gerin JL. Genotype-specific complementation of hepatitis delta virus RNA replication by hepatitis delta antigen. J Virol 1998; 72:2806-14.

7. Chang F-L, Chen P-J, Tu S-J et al. The large form of hepatitis Δ antigen is crucial for assembly of hepatitis Δ virus. Proc Natl Acad Sci USA 1991; 88:8490-8494.

8. Chang J, Morabda G, Taylor J. Limitations to replication of hepatitis delta virus in avian cells. J Virol 2000; 74:8861-8866.

9. Chang M-F, Baker SC, Soe LH et al. Human hepatitis delta antigen is a nuclear phosphoprotein with RNA-binding activity. J Virol 1988; 62:2403-2410.

10. Chang MF, Chen CH, Lin SL et al. Functional domains of delta antigens and viral RNA required for RNA packaging of hepatitis delta virus. J Virol 1995; 69:2508-2514.

11. Chang M F, Chen C-J, Chang S-C. Mutational analysis of delta antigen: Effect on assembly and replication of hepatitis delta virus. J Virol 1994; 68:646-653.

12. Chao M, Hsieh S-Y, Taylor J. The antigen of hepatitis delta virus: Examination of in vitro RNA binding specificity. J Virol 1991; 65:4057-4062.

13. Chao M, Hsieh S-Y, Taylor J. Role of two forms of hepatitis delta virus antigen: Evidence for a mechanism of self-limiting genome replication. J Virol 1990; 64:5066-5069.

14. Chen CW, Tsay YG, Wu HL et al. The double-stranded RNA-activated kinase, PKR, can phosphorylate hepatitis D virus small delta antigen at functional serine and threonine residues. J Biol Chem 2002; 277:33058-33067.

15. Chen P-J, Chang F-L, Wang C-J et al. Functional studies of hepatitis delta virus large antigen in packaging and replication inhibition: Role of the amino-terminal leucine zipper. J Virol 1992; 66:2853-2859.

16. Chen P-J, Kalpana G, Goldberg J et al. Structure and replication of the genome of hepatitis delta virus. Proc Natl Acad Sci USA 1986; 83:8774-8778.

17. Cheng Q, Jayan GC, Casey JL. Differential inhibition of RNA editing in hepatitis delta virus genotype III by the short and long forms of hepatitis delta antigen. J Virol 2003; 77:7786-95.

18. Choi S-S, Rasshofer R, Roggendorf M. Propagation of woodchuck hepatitis delta virus in primary woodchuck hepatocytes. Virology 1988; 167:451-457.

19. Chou H-C, Hsieh T-Y, Sheu G-T et al. Hepatitis delta antigen mediates the nuclear import of hepatitis delta virus RNA. J Virol 1998; 72:3684-3690.

20. Cole SM, Gowans EJ, Macnaughton TB et al. Direct evidence for cytotoxicity associated with expression of hepatitis delta virus antigen. Hepatology 1991; 13:845-851.

21. De Bruin W, Leenders W, Kos T et al. In vitro binding properties of the hepatitis delta antigens to the hepatitis B virus envelope proteins: Potential significance for the formation of delta particles. Virus Res 1994; 31:27-37.

22. Filipovska J, Konarska MM. Specific HDV RNA-templated transcription by Pol II in vitro. RNA 2000; 6:41-54.

23. Fu T-B, Taylor J. The RNAs of hepatitis delta virus are copied by RNA polymerase II in nuclear homogenates. J Virol 1993; 67:6965-6972.

24. Glenn JS, Watson JA, Havel CM et al. Identification of a prenylation site in delta virus large antigen. Science 1992; 256:1331-1333.

25. Goto T, Kato N, Ono-Nita SK et al. Large isoform of hepatitis delta antigen activates serum response factor-associated transcription. J Biol Chem 2000; 275:37311-6.

26. Guilhot S, Huang S-N, Xia Y-P et al. Expression of the hepatitis delta virus large and small antigens in transgenic mice. J Virol 1994; 68:1052-1058.

27. Hsieh S-Y, Chao M, Coates L et al. Hepatitis delta virus genome replication: A polyadenylated mRNA for delta antigen. J Virol 1990; 64:3192-3198.

28. Hsieh S-Y, Yang P-Y, Ou JT et al. Polyadenylation of the mRNA of hepatitis delta virus is dependent upon the structure of the nascent RNA and regulated by the small or large delta antigen. Nucleic Acids Res 1994; 22:391-396.

29. Huang WH, Yung BY, Syu WJ et al. The nucleolar phosphoprotein B23 interacts with hepatitis delta antigens and modulates the hepatitis delta virus RNA replication. J Biol Chem 2001; 276:25166-75.

30. Huang YH, Wu JC, Hsu SC et al. Varied immunity generated in mice by DNA vaccines with large and small hepatitis delta antigens. J Virol 2003; 77:12980-5.

31. Huang ZS, Wu HN. Identification and characterization of the RNA chaperone activity of hepatitis delta antigen peptides. J Biol Chem 1998; 273:26455-26461.

32. Hwang SB, Lai MMC. Isoprenylation masks a conformational epitope and enhances trans-dominant function of the large hepatitis delta antigen. J Virol 1994; 68:2958-2964.

33. Hwang SB, Lai MMC. Isoprenylation mediates direct protein-protein interactions between hepatitis large delta antigen and hepatitis B virus surface antigen. J Virol 1993; 67:7659-7662.

34. Hwang SB, Lee CZ, Lai MMC. Hepatitis delta antigen expressed by recombinant baculoviruses: Comparison of biochemical properties and post-translational modifications between the large and small forms. Virology 1992; 190:413-422.

34a. Hwang SB, Lai MMC. A unique conformation at the carboxyl terminus of the small hepatitis delta antigen revealed by a specific monoclonal antibody. Virology 1993; 193:924-931.

35. Jayan GC, Casey JL. Inhibition of hepatitis delta virus RNA editing by short inhibitory RNA-mediated knockdown of ADAR1 but not ADAR2 expression. J Virol 2002; 76:12399-404.

36. Jeng KS, Su PY, Lai MMC. Hepatitis delta antigens enhance the ribozyme activities of hepatitis delta virus RNA in vivo. J Virol 1996; 70:4205-4209.

37. Kuo MY-P, Chao M, Taylor J. Initiation of replication of the human hepatitis delta virus genome from cloned DNA: Role of Delta Antigen. J Virol 1989; 63:1945-1950.

38. Lazinski D, Grzadzielska E, Das A. Sequence-specific recognition of RNA hairpins by bacteriophage antiterminators requires a conserved arginine-rich motif. Cell 1989; 59:207-218.

39. Lazinski DW, Taylor JM. Expression of hepatitis delta virus RNA deletions: Cis and trans requirements for self-cleavage, Ligation and RNA packaging. J Virol 1994; 68:2879-2888.

40. Lazinski DW, Taylor JM. Relating structure to function in the hepatitis delta virus antigen. J Virol 1993; 67:2672-2680.

41. Lee CH, Chang SC, Chen CJ et al. The nucleolin binding activity of hepatitis delta antigen is associated with nucleolus targeting. J Biol Chem 1998; 273:7650-7656.

42. Lee CH, Chang SC, Wu CH et al. A Novel chromosome region maintenance 1-independent nuclear export signal of the large form of hepatitis delta antigen that is rerquired for the viral assembly. J Biol Chem 2001; 276:8142-8.

43. Lee C-Z, Chen PJ, Chen DS. Large hepatitis delta antigen in packaging and replication inhibition: Role of the carboxyl-terminal 19 amino acids and aminoterminal sequences. J Virol 1995; 69:5332-5336.

44. Lee C-Z, Chen P-J, Lai MMC et al. Isoprenylation of large hepatitis delta antigen is necessary but not sufficient for hepatitis delta virus assembly. Virology 1994; 199:169-175.

45. Lee C-Z, Lin J-H, Mcknight K et al. RNA-binding activity of hepatitis delta antigen involves two arginine-rich motifs and is required for hepatitis delta virus RNA replication. J Virol 1993; 67:2221-2229.

45a. Li YJ, Stallcup MR, Lai MMC. Hepatitis delta virus antigen is methylated at arginine residues, and methylation regulates subcellular localization and RNA replication. J Virol 2004; 78(23):13325-13334.

46. Lin J-H, Chang M-F, Baker SC et al. Characterization of hepatitis delta antigen: Specific binding to hepatitis delta virus RNA. J Virol 1990; 64:4051-4058.

47. Lo K, Hwang SB, Duncan R et al. Characterization of mRNA for hepatitis delta antigen: Exclusion of the full-length antigenomic RNA as an mRNA. Virology 1998a; 250:94-105.

48. Lo K, Sheu G-W, Lai MMC. Inhibition of cellular RNA polymerase II transcription by delta antigen of hepatitis delta virus. Virology 1998b; 247:178-188.

49. Luo G, Chao M, Hsieh SY et al. A specific base transition occurs on replicating hepatitis delta virus RNA. J Virol 1990; 64:1021-1027.

50. Macnaughton TB, Beard MR, Chao M et al. Endogenous promoters can direct the transcription of hepatitis delta virus RNA from a recircularized cDNA template. Virology 1993; 196:629-636.

51. Macnaughton TB, Gowans EJ, Jilbert AR et al. Hepatitis delta virus RNA, protein synthesis and associated cytotoxicity in a stably transfected cell line. Virology 1990a; 177:692-698.

52. Macnaughton TB, Gowans EJ, Mcnamara SP et al. Hepatitis δ antigen is necessary for access of hepatitis δ virus RNA to the cell transcriptional machinery but is not part of the transcriptional complex. Virology 1991; 184:387-390.

53. Macnaughton TB, Gowans EJ, Reinboth B et al. Stable expression of hepatitis delta virus antigen in a eukaryotic cell line. J Gen Virol 1990b; 71:1339-1345.

54. Macnaughton TB, Lai MMC. Genomic but not antigenomic hepatitis delta virus RNA is preferentially exported from the nucleus immediately after synthesis and processing. J Virol 2002a; 76:3928-3935.

55. Macnaughton TB, Lai MMC. Large hepatitis delta antigen is not a suppressor of hepatitis delta virus RNA synthesis once RNA replication is established. J Virol 2002b; 76:9910-9919.

56. Macnaughton TB, Li YI, Doughty AL et al. Hepatitis delta virus RNA encoding the large delta antigen cannot sustain replication due to rapid accumulation of mutations associated with RNA editing. J Virol 2003; 77:12048-12056.

57. Macnaughton TB, Shi ST, Modahl LE et al. Rolling circle replication of hepatitis delta virus RNA is carried out by two different cellular RNA polymerases. J Virol 2002; 76:3920-3927.

58. Modahl LE, Lai MM. The large delta antigen of hepatitis delta virus potently inhibits genomic but not antigenomic RNA synthesis: A mechanism enabling initiation of viral replication. J Virol 2000; 74:7375-7380.

59. Modahl LE, Lai MMC. Transcription of hepatitis delta antigen mRNA continues throughout hepatitis delta virus (HDV) replication: A new model of HDV RNA transcription and replication. J Virol 1998; 72:5449-5456.

60. Modahl LE, Macnaughton TB, Zhu N et al. RNA-dependent replication and transcription of hepatitis delta virus RNA involve distinct cellular RNA polymerases. Mol Cell Biol 2000; 20:6030-6039.

61. Mu J-J, Chen DS, Chen P-J. The conserved serine 177 in the delta antigen of hepatitis delta virus is one putative phosphorylation site and is required for efficient viral RNA replication. J Virol 2001; 75:9087-9095.

62. Mu J-J, Tsay YG, Juan LJ et al. The small delta antigen of hepatitis delta virus is an acetylated protein and acetylation of lysine 72 may influence its cellular localization and viral RNA synthesis. Virology 2004; 319:60-70.

63. Mu J-J, Wu H-L, Chiang B-L et al. Characterization of the phosphorylated forms and the phosphorylated residues of hepatitis delta virus delta antigens. J Virol 1999; 73:10540-10545.

64. Otto JC, Casey PJ. The hepatitis delta virus large antigen is farnesylated both in vitro and in animal cells. J Biol Chem 1996; 271:4569-4572.

65. Poisson F, Roingeard P, Baillou A et al. Characterization of RNA-binding domains of hepatitis delta antigen. J Gen Virol 1993; 74:2473-2477.

66. Polo JM, Jeng KS, Lim B et al. Transgenic mice support replication of hepatitis delta virus RNA in multiple tissues, particularly in skeletal muscle. J Virol 1995; 69:4880-4887.

67. Polson AG, Bass BL, Casey JL. RNA editing of hepatitis delta virus antigenome by dsRNA- adenosine deaminase. Nature (London) 1996; 380:454-456.

68. Rizzetto M, Canese MG, Arico S et al. Immunofluorescence detection of A new antigen-antibody system (Delta/Anti-Delta) associated with hepatitis B virus in liver and serum of HBsAg carrier. Gut 1977; 18:997-1003.

69. Rizzetto M, Hoyer B, Canese MG et al. Delta agent: Association of Δ antigen with hepatitis B surface antigen and RNA in serum of Δ-infected chimpanzees. Proc Natl Acad Sci USA 1980; 77:6124-6128.

70. Rozzelle JE, Wang JG, Wagner DS et al. Self-association of a synthetic peptide from the N terminus of the hepatitis delta virus protein into an immunoreactive alpha helical multimer. Proc Natl Acad Sci USA 1995; 92:382-386.

71. Ryu W-S, Bayer M, Taylor J. Assembly of hepatitis delta virus particles. J Virol 1992; 66:2310-2315.

72. Ryu W-S, Netter H J, Bayer M et al. Ribonucleoprotein complexes of hepatitis delta virus. J Virol 1993; 67:3281-3287.

73. Sheu G-T, Lai MMC. Recombinant hepatitis delta antigen from E. Coli promotes hepatitis delta virus RNA replication only from the genomic strand but the antigenomic strand. Virology 2000; 278:578-586.

74. Sureau C, Jacob JR, Eichberg JW et al. Tissue culture system for infection with human hepatitis delta virus. J Virol 1991; 65:3443-3450.

75. Sureau C, Moriarty AM, Thornton GB et al. Production of infectious hepatitis delta virus in vitro and neutralization with antibodies directed against hepatitis B virus preS antigens. J Virol 1992; 66:1241-1245.

76. Tai F-P, Chen P-J, Chang F-L et al. Hepatitis delta virus cDNA can be used in transfection experiments to initiate Viral RNA replication. Virology 1993; 197:137-142.

77. Tavanez JP, Cunha C, Silva MCA et al. Hepatitis delta virus ribonucleoproteins shuttle between the nucleus and the cytoplasm. RNA 2002; 8:637-646.

78. Taylor J, Mason W, Summers J et al. Replication of human hepatitis delta virus in primary cultures of woodchuck hepatocytes. J Virol 1987; 61:2891-2895.

79. Taylor JM. Human hepatitis delta virus: An agent with similarities to certain satellite RNAs of plants. Curr Top Microbiol Immunol 1999; 239:107-122.

80. Wang CC, Chang TC, Lin CW et al. Nucleic acid binding properties of the nucleic acid chaperone domain of hepatitis delta antigen. Nucleic Acids Res 2003; 31:6481-6492.

81. Wang D, Pearlberg J, Liu YT et al. Deleterious effects of hepatitis delta virus replication on host cell proliferation. J Virol 2001; 75:3600-4.

82. Wang H-W, Chen P-J, Lee C-Z et al. Packaging of hepatitis delta virus RNA via the RNA-binding domain of hepatitis delta antigens: Different roles for the small and large delta antigens. J Virol 1994; 68:6363-6371.

83. Wang J-G, Lemon SM. Hepatitis delta virus antigen forms dimers and multimeric complexes in vivo. J Virol 1993; 67:446-454.

84. Wei Y, Ganem D. Activation of heterologous gene expression by the large isoform of hepatitis delta antigen. J Virol 1998; 72:2089-2096.

85. Weiner AJ, Choo Q-L, Wang K-S et al. A single antigenomic open reading frame of the hepatitis delta virus encodes the epitope(s) of both hepatitis delta antigen polypeptides P24$^\delta$ and P27$^\delta$. J Virol 1988; 62:594-599.

86. Wong SK, Lazinski DW. Replicating hepatitis delta virus RNA is edited in the nucleus by the small form of ADAR1. Proc Natl Acad Sci USA 2002; 99:15118-23.

87. Xia Y-P, Chang M-F, Wei D et al. Heterogeneity of hepatitis delta antigen. Virology 1990; 178:331-336.

88. Xia Y-P, Lai MMC. Oligomerization of hepatitis delta antigen is required for both the trans-activating and trans-dominant inhibitory activities of the delta antigen. J Virol 1992; 66:6641-6648.

89. Xia Y-P, Yeh C-T, Ou J-H et al. Characterization of nuclear targeting signal of hepatitis delta antigen: Nuclear transport as a protein complex. J Virol 1992; 66:914-921.

90. Yamaguchi Y, Delehouzee S, Handa H. HIV and Hepatitis delta virus: Evolution takes different paths to relieve blocks in transcriptional elongation. Microbes Infect 2002; 4:1169-1175.

91. Yamaguchi Y, Filipovska J, Yano K et al. Stimulation of RNA polymerase II elongation by hepatitis delta antigen. Science 2001; 293:124-127.

92. Yeh TS, Lo SJ, Chen PJ et al. Casein kinase II and protein kinase C modulate hepatitis delta virus RNA replication but not empty viral particle assembly. J Virol 1996; 70:6190-6198.

93. Zuccola HJ, Rozzelle JE, Lemon SM et al. Structural basis of the oligomerization of hepatitis delta antigen. Structure 1998; 6:821-830.

CHAPTER 5

Hepatitis Delta Virus RNA Editing

John L. Casey*

Summary

The genome of hepatitis delta virus (HDV) is the smallest known to infect man. Encoding just one protein, hepatitis delta antigen (HDAg), HDV relies heavily on host functions and on structural features of the viral RNA. A good example of this reliance is found in the process known as HDV RNA editing, which requires particular structural features in the HDV antigenome, and a host RNA editing enzyme, ADAR1. During replication, the adenosine in the amber stop codon in the viral gene for the short form of HDAg (HDAg-S) is edited to inosine. As a result, the amber stop codon in the HDAg-S open reading frame is changed to a tryptophan codon; the reading frame is thus extended by 19 or 20 codons and the longer form of HDAg, HDAg-L, is produced. This change serves a critical purpose in the HDV replication cycle because HDAg-S supports viral RNA replication, while HDAg-L is required for virion packaging but inhibits viral RNA replication. This review will cover the mechanisms of RNA editing in the HDV replication cycle and the regulatory mechanisms by which HDV controls editing.

What Is RNA Editing?

RNA editing can be loosely defined as the site-specific modification of an RNA sequence from that of its template by mechanisms other than splicing. The term was first used in the late 1980's to describe an unusual process in which multiple U's are inserted and deleted in trypanosome mitochondrial mRNAs.[1] As a result of the insertions/deletions, the coding capacity of the affected mRNAs is dramatically altered. The usage of the term was subsequently expanded as it was applied to other examples of nucleotide changes in mRNA that changed the coding capacity, including deamination of C to U in apoB100 mRNA in small intestine,[2] deamination of A to I in glutamate receptor subunit B (gluRB) premRNA in brain,[3] and insertion of nontemplated G's in the P gene of paramyxoviruses.[4] While collectively referred to as RNA editing, these sequence revisions involve a wide range of mechanisms. In the two types of editing used by mammalian cells, C to U and A to I, the modified base within the RNA molecule is deaminated and there is no evidence that phosphate backbone is broken during the editing process.

The type of RNA editing used by HDV is adenosine deamination. In this process, the amino group of adenosine is removed and replaced with a keto oxygen. Because this position of

*John L. Casey—Department of Microbiology and Immunology, Georgetown University Medical Center, 3900 Reservoir Rd., NW, Washington, District of Columbia, 20007 U.S.A. Email:caseyj@georgetown.edu

Hepatitis Delta Virus, edited by Hiroshi Handa and Yuki Yamaguchi.
©2006 Landes Bioscience and Springer Science+Business Media.

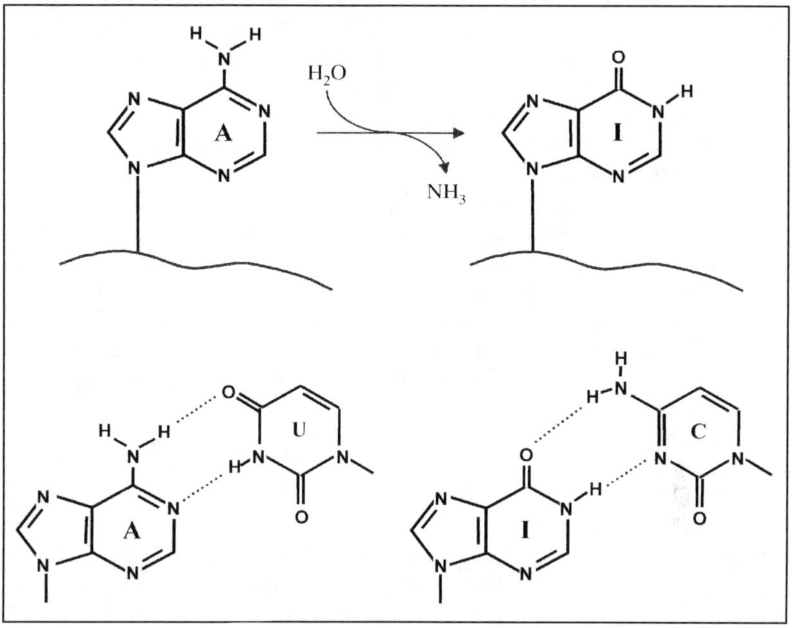

Figure 1. Adenosine deamination. The upper panel shows the replacement of the amino group of adenosine by oxygen to generate inosine. The horizontal line indicates the RNA phosphate backbone, which is not broken during the deamination reaction. The lower panel shows the effects of deamination on base-pairing. Adenosine forms Watson-Crick base-pairs with uracil; whereas inosine base-pairs with cytidine. Hydrogen bonds are designated by dotted lines.

the base is changed from a hydrogen bond donor to an acceptor, the Watson-Crick base-pairing preference of this nucleotide is changed from pairing with U to pairing with C (Fig. 1). Therefore, in any subsequent functions that involve base-pairing (such as translation, RNA-templated transcription, and splice site identification) the edited position will behave as G rather than the original A. Editing has the potential to produce as many as 15 different recodings of an RNA transcript, including the creation of a methionine start codon and the abolition of stop codons. Thus, for example, when the adenosine at the R/G site in the glutamate receptor subunit B mRNA is edited, a CAG arginine codon is changed to CIG, which behaves like CGG, and encodes glycine; as a result of this change, the cation permeability of glutamate receptor channels in mammalian brain are changed. As indicated by this example, sites in RNAs that undergo adenosine deamination have been named according to the coding change brought about by editing. Thus, because editing on the HDV RNA changes an amber stop codon to a tryptophan (W) codon, the position on the HDV RNA at which editing occurs is called the amber/W site.

Mechanism of HDV RNA Editing

HDV Produces Two Forms of HDAg from the Same Gene
Early analyses of HDV proteins showed that there are two electrophoretic forms of HDAg, but it was not clear how these forms differed biochemically and functionally.[5-8] (These forms were sometimes referred to by their apparent molecular weights, p-24 and p-27; they are denoted here as HDAg-S and HDAg-L for short and long, respectively.) Following the cloning of HDV cDNAs,[9,10] a series of studies illuminated the functional roles of HDAg-S and HDAg-L

in HDV replication. Taylor's group showed that while HDAg-S is required for replication of HDV RNA, HDAg-L inhibits replication.[11,12] Subsequent studies from several laboratories showed that HDAg-L interacts with the envelope protein of the helper virus, HBV, and is required for the formation of HDV particles.[13-15]

Despite these advances, the mechanisms leading to the formation of HDAg-S and HDAg-L were not clear. Cloning and sequencing of the genome in 1986 indicated heterogeneity at several positions in the 1679 nt genome;[9] most of this heterogeneity involved single base transitions (A vs. G and C vs. U). Although the significance was not fully understood at the time, heterogeneity at one position, in particular, proved to be important. This position is near the end of the single conserved open reading frame, which is in the antigenome sequence (HDV is a negative strand RNA virus). In some clones this position was found to be adenosine, which is part of the UAG (amber) stop codon of HDAg-S; in other clones it was found to be guanosine, which changes the codon to UGG (W, tryptophan) and thereby extends the open reading frame by 19 amino acids.[9,16] Expression of protein from clones that contained either the UAG or UGG sequence showed that the former encoded HDAg-S and the latter HDAg-L.[16,17] Still, it was not clear how this heterogeneity fit into the HDV replication scheme.

An important advance came when HDV infection was initiated in chimpanzees by transfection with a cloned HDV cDNA expression construct that could initiate HDV replication.[18] Remarkably, while the transfected genome encoded only HDAg-S, both HDAg-S and HDAg-L were detected in the liver and in HDV particles isolated from the serum of the infected chimpanzee.[18] This result was further extended by analysis of HDV replication in cultured cells: careful examination of western blots indicated that HDAg-L also became detectable several days after transfection of cultured cells with a cDNA clone that initiated HDV RNA replication and that encoded only HDAg-S.[19] No HDAg-L was detected when cells were transfected with an expression construct for HDAg-S that did not produce replicating HDV RNA. Thus, the appearance of HDAg-L was linked to HDV replication.

Analysis of HDV RNA isolated from the serum of the transfected chimpanzee and from transfected cultured cells showed that heterogeneity appeared at the position corresponding to the adenosine in the UAG stop codon for HDAg-S.[19] This finding recalled the heterogeneity observed at this position during the sequencing of the original isolates of HDV and led to the suggestion that this heterogeneity was responsible for the appearance of HDAg-L. Thus, during the course of HDV replication, some genomes encoding HDAg-S are converted, or edited, to encode HDAg-L. Because of the different functions of HDAg-S and HDAg-L, editing is part of a classic switch from viral RNA replication to packaging, and plays a central role in the HDV replication cycle.

HDV Antigenomic RNA Is Edited by the Host RNA Adenosine Deaminase ADAR1

One difficulty encountered in establishing the mechanism of editing at the amber/W site was identifying the RNA substrate: assays performed on replicating RNAs could not definitively determine whether the substrate for editing was the genome or the antigenome, or even whether editing was the result of cotranscriptional misincorporation. Initial attempts led to the erroneous suggestion that the genomic RNA might be the substrate, in which case editing would occur as a U to C transition.[20,21] However, the creation of nonreplicating RNA expression constructs that could exclusively produce either genomic or antigenomic RNA in transfected cells led to the unambiguous conclusion that editing occurs on the antigenome RNA.[22] This result was further supported by analysis of editing on in vitro-transcribed RNAs mixed with nuclear extracts: only antigenomic RNA was edited at the amber/W site.[22] This observation indicated that HDV editing occurs post-transcriptionally, and is not the result of transcriptional misincorporation.

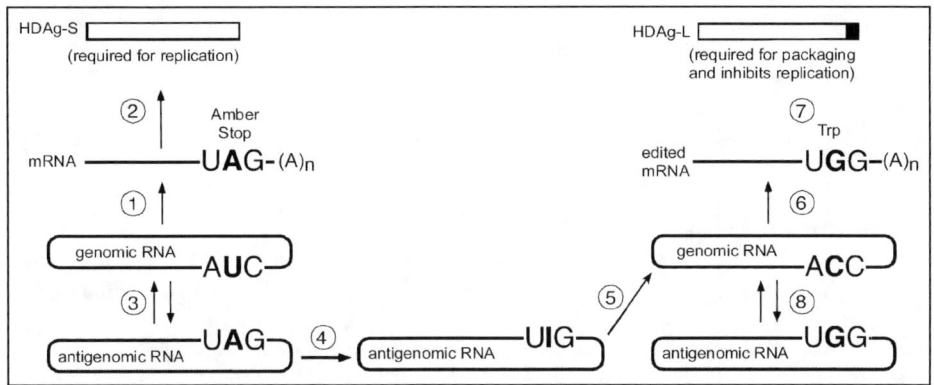

Figure 2. The role of RNA editing in the HDV replication cycle. (Adapted from Polson et al 1996.) 1) Synthesis of mRNA encoding HDAg-S; 2) translation of HDAg-S, which is required for RNA replication; 3) replication of full-length antigenomic and genomic RNA; 4) During replication some of the antigenomic RNA is edited at the amber/W site by the host RNA adenosine deaminase ADAR1; 5) antigenomic RNA containing I at the editing site serves as template for the synthesis of genomic RNA containing C at the complementary position; 6) synthesis of mRNA encoding HDAg-L; 7) translation of HDAg-L, which inhibits RNA replication, and is required for packaging; 8) replication of genomic and antigenomic RNA encoding HDAg-L.

Because HDAg is encoded on the HDV antigenome (HDV is a negative strand RNA virus) editing on the antigenome is consistent with A to I editing by RNA adenosine deaminase,[19] an enzymatic activity which is present in nuclear extracts from numerous metazoan species and can edit adenosines in double-stranded RNAs. Subsequently, it was shown that RNA adenosine deaminase (ADAR) from *Xenopus laevis* can edit the amber/W site in the HDV antigenome with considerable specificity in vitro.[23] Because the HDV amber/W site was edited with high specificity in vitro using just purified ADAR, no additional factors aside from HDV RNA and RNA adenosine deaminase are required for amber/W site editing to occur.

In mammalian cells two related genes, ADAR1 and ADAR2, have been identified that encode proteins capable of editing adenosine in RNA.[24-27] These proteins contain a catalytic deaminase domain along with 3 or 2, respectively, copies of dsRNA binding motifs (DRBMs). Both genes are essential for viability in mice.[28,29] Substrates for ADAR1 and ADAR2 include a number of host premRNAs,[30,31] including the glutamate receptor subunit B (gluR-B) Q/R and R/G sites, the gluR-5 and gluR-6 Q/R sites, a splice site in ADAR2, and several sites in the serotonin receptor $5HT_{2C}R$.[31] Both ADAR1 and ADAR2 can edit HDV RNA at the amber/W site in transfected cultured cells.[32-34] While ADAR1 and ADAR2 are expressed in a variety of tissues, levels of ADAR2 are highest in brain, and the level of ADAR1 mRNA expression is considerably higher than ADAR2 in liver. Knockdown experiments using siRNA have shown that the short form of ADAR1, which is localized in the nucleus, is responsible for amber/W site editing during HDV replication.[35,36]

In the current model for RNA editing in the HDV replication cycle (Fig. 2), the amber/W site in full-length antigenome RNA is edited from A to I by the host RNA adenosine deaminase ADAR1. The edited antigenome serves as template for the synthesis of genome RNA that then contains C at the position complementary to the amber/W site in the antigenome. This genome subsequently serves as template for transcription of mRNA that contains the UGG tryptophan codon and therefore encodes HDAg-L. It is important to note that, in this model, HDV mRNA is not edited directly, unlike cellular mRNA substrates for RNA adenosine deamination. Rather, editing occurs on the full-length antigenome, which is a replication intermediate. Consistent with this model, sequences required for forming the structure required for editing (see below) are more than 300nt downstream of the polyadenylation and ribozyme sites, and are not included in the mRNA sequence. Furthermore, analysis of RNA in viral particles indicates that genome RNAs contain the expected C at the position complementary to the amber/W site.

Regulation of HDV RNA Editing

As a result of the scheme shown in Figure 2, editing accumulates in both antigenome and genome RNA and editing levels in HDV RNA within an infected cell at any given time represent the integration (in the mathematical sense) of all editing events within that cell up to that time. The cost of this mechanism to the virus is that a fraction of viral particles contain genomes that encode HDAg-L, which does not support HDV RNA replication; such particles are therefore not likely to be infectious. Thus, HDV must control the level of editing in order to ensure viability.

Regulation of HDV RNA editing occurs on several levels. First, the HDV antigenome contains about 337 adenosines, but editing is highly specific for the amber/W site. Second, both the rate and extent of editing appear to be carefully controlled. Some host substrates for editing exhibit modification rates approaching 100%, and this editing likely occurs rapidly because all known host substrates are premRNAs that are edited prior to splicing. In contrast, for HDV, edited viral RNAs accumulate slowly and typically level off at levels less than 30% after 12 days in transfected cells in culture.

Cis Elements Required for Editing

ADAR1 and ADAR2 exhibit distinct but overlapping substrate specificities. Although some substrates, such as the gluRB Q/R site, show a clear preference for one ADAR, many substrates, including the HDV amber/W site, can be edited by both. Inspection of the predicted structures of known editing sites reveals several common features (Fig. 3). All substrates include at least 6 contiguous base pairs around the editing site, and many substrates contain more. Base-pairing on the 5' side of sites varies between 2 and 5 base-pairs. On the 3' side base-pairing is greater, in most cases extending for at least about 20 base-pairs that are disrupted by mismatches, bulges and internal loops. The role of these disruptions may be to position the ADAR protein via the double-stranded RNA binding domains such that the deaminase domain is positioned correctly at the editing site.[31,37-39] In an apparent contradiction to this idea, some studies have suggested that extensive base-pairing 3' of editing sites may not be essential for efficient editing. Sato and Lazinski[34] showed that a minimal substrate that was derived from the HDV amber/W site and that contained only 8 base-pairs could be efficiently edited, and Herbert et al[40] showed that ADAR1 could efficiently edit even when the three DRBMs were removed. Both of these studies examined editing in transfected cells expressing high levels of ADAR, and could indicate that the deaminase domain itself possesses some RNA binding activity that can be effective under certain conditions.

It is interesting to note that the base-pairing on the 3' side of the HDV amber/W site is more frequently disrupted by bulges and mismatches than other editing substrates. Mutations that improve base-pairing in this region in the HDV genotype I RNA lead to increased editing and markedly decreased viral RNA replication (Sato and Lazinski, personal communication; Jayan and Casey, unpublished data). There are two likely reasons for the presence of these disruptions 3' of the HDV amber/W site. First, the more extensive base-pairing found in other editing substrates could interfere with HDV RNA replication (either because RNA structures required for viral RNA replication cannot tolerate extensive base-pairing, or because the base-pairing may trigger cellular antiviral responses to dsRNA segments). Second, the

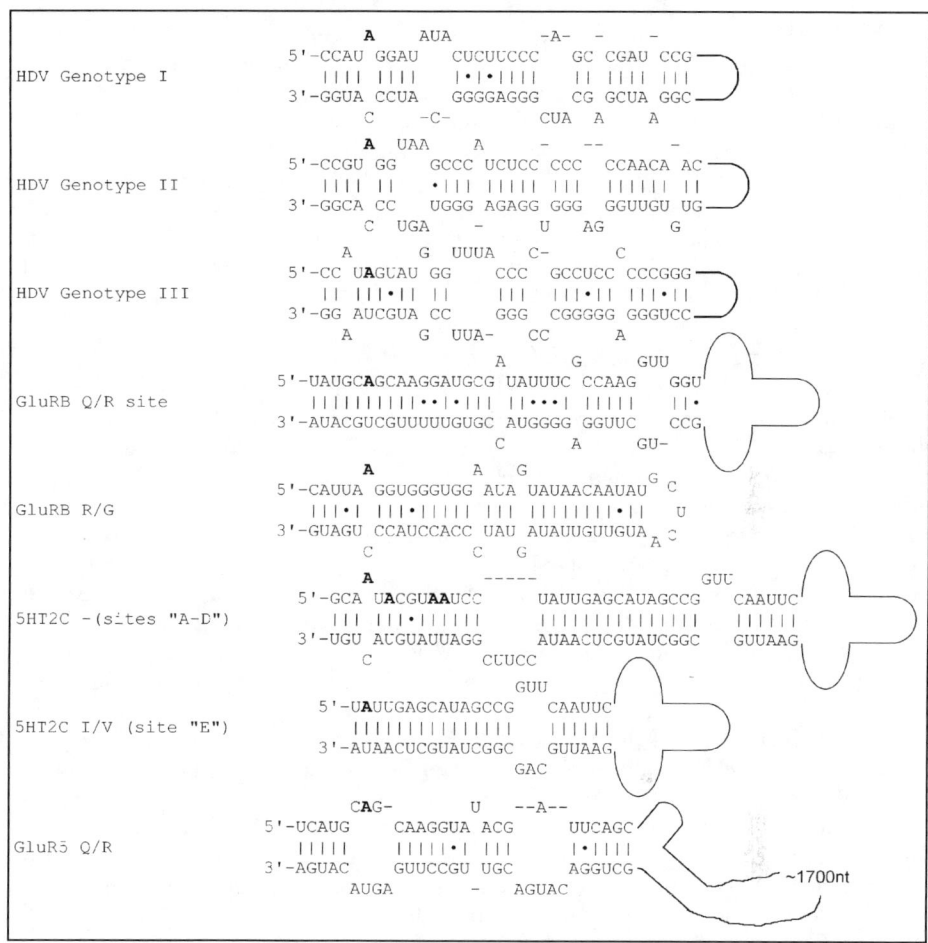

Figure 3. Predicted RNA secondary structure around known sites for RNA adenosine deamination. The structure shown for the gluRB Q/R site was obtained using the RNA secondary structure prediction program mfold,[56] and is different from another proposed structure.[57] Vertical lines indicate A-U and G-C base-pairs. Dots indicate G-U pairs.

disruptions may be important for preventing excessive editing (Lazinski and Sato, personal communication; Jayan and Casey, unpublished data); that is, the HDV editing site (at least for genotype I) may be selected to be sub-optimal in order to prevent the rapid accumulation of too much HDAg-L, which would inhibit viral RNA replication.

The structure in the immediate vicinity of edited adenosines varies somewhat, but in most substrates the target adenosine occurs as either an A-U pair or an A-C mismatch pair. Mutational analysis of some substrates, including the HDV genotype I amber/W site, indicates that editing levels are much higher when the adenosine occurs as an A-C mismatch rather than an A-U pair.[20,23,33,41,42] Moreover, at least for HDV genotype I, any change in the position opposite the amber/W site (deletion, or substitution by A or G), led to markedly reduced editing levels.[20]

Despite recognition of the common features among editing sites noted above, the sequence and structural determinants for highly specific editing are still not well understood.

Only a handful of substrates for highly specific editing have been identified to date in mammals, and it is anticipated that many more remain to be identified. Knowledge of sequence and structural requirements for editing will likely facilitate the prediction of potential adenosine deamination editing sites from analysis of genomic sequences. Moreover, it is reasonable to expect that differences in editing levels among different substrate adenosines are due in part to variations in the structure of the RNA in the vicinity of the editing sites. Understanding the effects of structural variations will contribute to our understanding of how this important posttranscriptional regulatory mechanism is modulated. Thus, defining the structural determinants for editing remains an important goal; it is likely that the HDV amber/W site will be a useful tool in this endeavor.

Genotype Variations and Amber/W Editing

The sequence/structures around the amber/W site and the C-terminal sequences specific to HDAg-L are distinguishing features of the HDV genotypes.[43-48] The genotype-specific variations in these two functional elements may be the result of the connection between amber/W editing and HDAg-L function, and might reflect different requirements for HDAg-L during the course of replication in the three genotypes.

In HDV genotype I the structure required for editing at the amber/W site is part of the unbranched rod structure characteristic of HDV RNA.[20] The 8 Watson-Crick base-pairs flanking the amber/W site and the A-C mismatch pair involving the amber/W site are highly conserved among over 50 genotype I sequences.[44,49] Indeed, the only variation is the occasional substitution of the A-U pair 2 positions 5' of the amber/W site by a G-C pair. Site-directed mutagenesis studies have shown that both the base-pairing and the A-C mismatch are required for editing.[20]

In the unbranched rod structure formed by HDV genotype III RNA the base-pairing in the immediate vicinity of the amber/W site is disrupted such that this structure does not function as a substrate for amber/W editing.[45] Rather, editing in genotype III requires an alternative structure that creates better base-pairing in the immediate vicinity of the amber/W site (Fig. 4).[45] Remarkably, this structure, termed the double hairpin, differs from the unbranched rod structure by nearly 80 base-pairs. The mechanisms by which the genotype III double hairpin RNA structure is formed and rearranges to the unbranched rod (which is required for RNA replication)[45] remain to be determined, but it appears that, as for viroid RNAs, RNA structural dynamics play an important role in HDV replication, not only in the activity of the highly structured ribozyme, but in other processes as well.

The structure required for editing in HDV genotype II has not yet been determined. However, comparative analysis of the predicted secondary structure in the vicinity of the genotype II amber/W site in the unbranched rod reveals a conserved structure more similar to the genotype I structure than that in the genotype III unbranched rod, but still more disrupted than the genotype I structure.[46,48] This slightly disrupted structure, if it is indeed used as the substrate for genotype II amber/W editing, may be consistent with the observation that in transfected cells, editing levels were lower for replicating genotype II RNA than for genotype I RNA.[46]

Inspection of the predicted structures around the amber/W sites in the three genotypes indicates that the genotype II and III amber/W sites vary from the type I site at positions that have been shown to be essential for editing in type I. For example, the A-C mismatch pair that involves the amber/W adenosine and which is highly conserved among genotype I isolates, occurs as an A-U pair in genotype III; when introduced by site-specific mutagenesis into a genotype I genome, this specific change substantially reduces editing and virus production.[20,50] Perhaps the variations at the genotype II and III sites can be explained by compensatory effects, such as changes elsewhere in the editing site region, including sequences/structures 3' of the

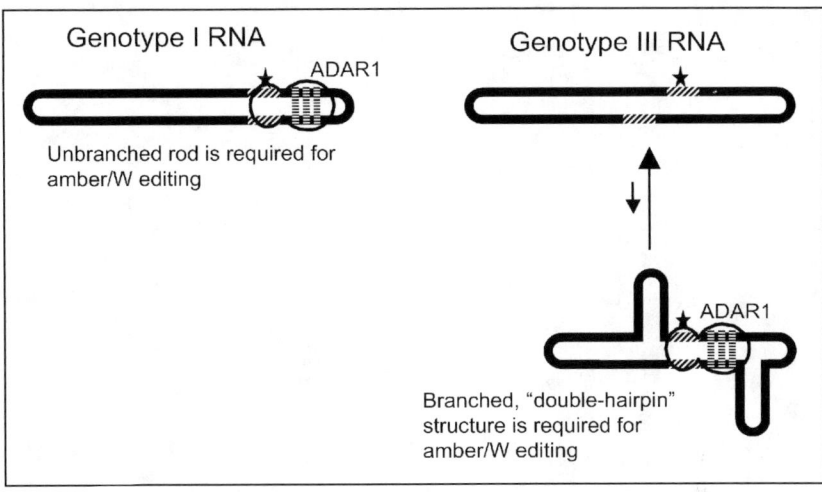

Figure 4. Schematic of RNA structures required for amber/W site editing in HDV genotypes I and III. The location of the amber/W site is indicated by the star. Sequences involved in forming the base-paired secondary structure essential for editing are indicated by diagonal hashes. In the genotype I unbranched rod structure these sequences are correctly positioned to form the structure required for editing,[20] while in genotype III they are not.[45] For genotype III the structure required for editing contains two ca. 25 nt stem-loops and deviates from the unbranched rod by about 80 base-pairs. The vertical arrows indicate that the distribution of the genotype III RNA between these two structures favors the unbranched rod structure.[45,54] Editing levels in HDV genotype III are restricted by this distribution. The three double-stranded RNA binding domains in ADAR1 are indicated by striped bars.

editing site, or differences in the mechanisms/processes by which HDV regulates editing during replication. Variations among the amber/W sites in the three HDV genotypes may provide a valuable opportunity for analyzing the structural determinants for RNA editing and evaluating the effects of different sequences/structures on editing at the HDV amber/W site.

Specificity of Editing

ADAR1 and ADAR2 can extensively edit long ($\geq$ 50 base-pairs) double-stranded RNAs, in which up to 50% of adenosines may be deaminated. The role of this activity in cells is not clear, but the fact that dsRNA is a target and that one form of ADAR1 is induced by interferon[27] has led to the suggestion that editing of dsRNA may be part of the cellular response to virus infection. Clearly promiscuous editing such as occurs on dsRNA could be deleterious to virus replication. Indeed, spurious editing on HDV RNA by overexpressed ADAR1 and ADAR2 led to the production of protein variants that inhibited replication.[32] Whether interferon treatment increases editing at the amber/W site, or elsewhere on the HDV RNA, remains to be determined.

Even though HDV RNA exhibits significant base-pairing in the unbranched rod structure, promiscuous editing does not typically occur during HDV infection; the amber/W site is edited 600-fold more efficiently than the other 337 adenosines in the RNA.[51] It is likely that the primary and secondary structure of the HDV RNA have evolved to avoid undesirable (for the virus) editing at sites other than amber/W. Guanosine is by far the most common 5' neighbor for adenosine in both the HDV genome and antigenome, and the ratios of observed to expected occurrences for the dinucleotides GA and UC (which would be GA in the complementary strand) are higher than for any other dinucleotides (Table 1). This bias may be due, in part,

Table 1. Dinucleotide frequencies in HDV RNA[a]

Dinucleotide	No. of Occurences	Observed/ Expected
GA	161	1.65
TC	160	1.54
AG	134	1.38
CT	140	1.35
CC	204	1.28
GG	177	1.24
AA	79	1.19
TT	79	1.17
AT	56	0.83
CG	112	0.74
TG	66	0.67
AC	65	0.63
GT	62	0.63
CA	62	0.6
GC	89	0.59
TA	32	0.48

[a] based on HDV genotype I prototype, accession no. X04451

to selection for sequences that place non-amber/W adenosines in contexts that are less likely to be edited: analysis of editing on dsRNAs has indicated that adenosines flanked by a 5' guanosine are much less likely to be deaminated than other adenosines.[38] As for secondary structure, base-pairing in the HDV RNA unbranched rod structure is interrupted by frequent bulges, internal loops and mismatches, which have been shown to restrict editing on artificial dsRNA substrates.[37,39,52]

Regulation of Editing

HDV must regulate both the rate and the extent of editing at the amber/W site because HDAg-L, which is produced as a result of editing, is necessary for virion production but inhibits viral RNA replication. Varying the efficiency of editing at the amber/W site, either by altering levels of ADAR expression or by the introduction of mutations near the amber/W site, can affect HDV replication, virus production, or both.[32,45] Premature editing at the amber/W site results in reduced levels of RNA replication and reduced production of viable virions because edited antigenomes encode HDAg-L, which is a trans-dominant inhibitor of HDV RNA replication.[12,50,53] Insufficient editing can lead to increased intracellular HDV RNA replication, but inhibits virion production.[45] In addition to the rate, the extent of editing must also be controlled because the mechanism of editing in the HDV replication scheme (Fig. 2) produces genomes encoding HDAg-L. These genomes are packaged but are not likely to be infectious because HDAg-L does not support HDV RNA replication. Thus, the kinetics and extent of editing are likely regulated during HDV replication to maximize the rate and amount of infectious virus produced.

Control mechanisms for editing rely on several viral components and functions, including: RNA structure, HDAg, and viral RNA replication. HDV does not appear to regulate editing by affecting ADAR1 expression because ADAR1 levels are unaffected by HDV replication.[35] Some of the control mechanisms may be described as passive, in that they are not affected by (or responsive to) the level of editing. This category includes the secondary structure

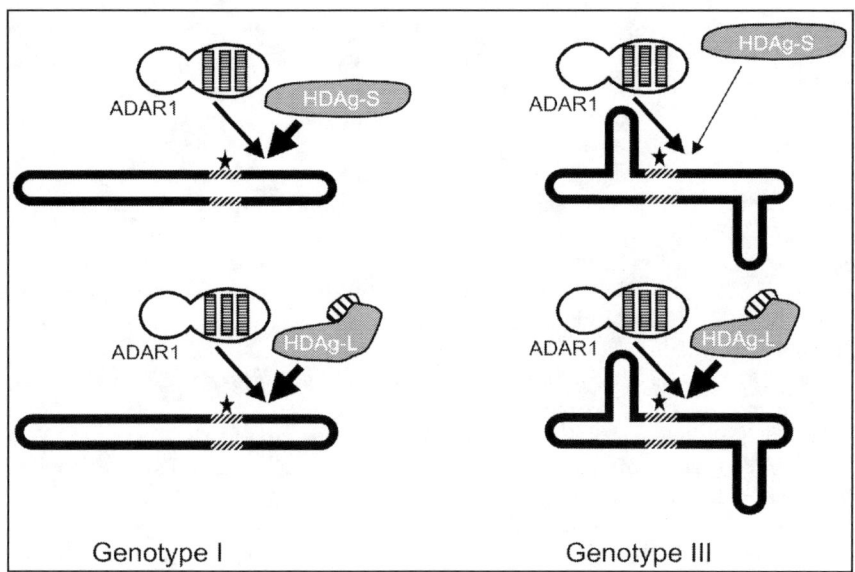

Figure 5. Schematic of the regulation of editing in HDV genotype I and III by HDAg. Left) genotype I. Both HDAg-S and HDAg-L effectively inhibit editing, most likely by binding the RNA near the editing site and limiting access of ADAR1. Right) genotype III. HDAg-L, but not HDAg-S can effectively inhibit editing. HDAg-L is drawn with the C-terminal extension indicated by the small segment with diagonal hashing

of the RNA around the amber/W site. As mentioned above, the disruptions in base-pairing 3' of the amber/W site in HDV genotype I create a sub-optimal substrate for editing. Mutations that increase base-pairing in this region increase editing, but severely reduce replication and virion production (Sato and Lazinski, personal communication; Jayan and Casey, unpublished). It is not yet known whether the structures in the vicinity of the amber/W sites of genotypes II and III are also sub-optimal. One potential dilemma for the virus that is posed by using a sub-optimal structure to limit editing efficiency is that the specificity of editing is likely to be compromised because the specificity is determined by the ratio of the efficiency of editing at the amber/W site to the efficiency of editing at other "nonspecific" sites. The danger for the virus of nonspecific editing is the production of additional genomes defective for replication, or even the creation of dominant negative HDAg-S mutants.[32] Thus, there may be limits as to how much amber/W editing can be restricted by using sub-optimal structures. HDV does appear to have a mechanism for minimizing the effects of editing at nonamber/W sites: in one study of HDV replicating HDV in transfected cells, all nonamber/W changes that occurred during replication were found on genomes that were also edited at the amber/W site.[51]

HDV genotype I uses an additional mechanism to slow down editing early in the replication cycle. For this genotype, HDAg-S is a strong inhibitor of editing (Fig. 5). While editing on replicating RNA 2-3 days post-transfection is nearly undetectable, up to 40% of nonreplicating RNAs produced in transfected cells in the absence of HDAg are edited. However, cotransfection of an HDAg-S expression construct leads to markedly reduced levels of editing on nonreplicating RNAs, most likely by binding to HDV RNA and preventing access of ADAR1.[51] The levels of HDAg-S required for this inhibition are similar to those seen in cells replicating HDV RNA. Thus, it appears that HDAg-S prevents the rapid accumulation of editing early in the HDV genotype I replication cycle.

On the other hand, for HDV genotype III, HDAg-S is not an effective inhibitor of editing and likely does not play a direct role in limiting editing levels.[54] Rather, HDV genotype III uses the distribution of the RNA between at least two conformations to restrict editing.[45,54] Only RNA molecules that adopt the double hairpin structure can be edited (Fig. 4). However, the majority of the genotype III RNA appears to assume the unbranched rod conformation, which is not a substrate for editing.[45,54] Thus, while the amber/W site itself in genotype III RNA can be edited with efficiency similar to the genotype I site, editing levels in nonreplicating genotype III RNAs are much lower because most of the RNA assumes the unbranched rod conformation, which is not a substrate for editing.[45] The introduction of mutations in the genotype III RNA that shift the distribution of the RNA to the double hairpin structure increases editing to levels comparable with those seen with nonreplicating genotype I RNA.[45]

It is not clear whether the inhibition of editing in genotype I or the conformational control of editing in genotype III is influenced by levels of editing and/or replication. Possibly, the HDAg:RNA ratio may vary during replication and thereby affect HDV genotype I editing rates. Likewise, for genotype III, if the RNA transcription rate varies during HDV genotype III RNA replication, such variations could affect editing by altering the distribution of the RNA between the double hairpin and unbranched rod structures.

Other mechanisms to control editing are by their nature responsive to editing levels. In HDV genotype III, HDAg-L is a much better inhibitor of editing than is HDAg-S, and is likely involved in a negative feedback loop to limit editing levels (Fig. 5).[54] This regulatory behavior requires the double hairpin structure peculiar to HDV genotype III amber/W editing. The reason for the differential effects of HDAg-S and HDAg-L is not yet clear; the hairpin on the 3' side of the amber/W site plays an essential role,[54] and could interfere with HDAg-S binding near the amber/W site. Genotype I does not use the same mechanism to control editing levels because genotype I HDAg-S and HDAg-L do not exhibit differential effects on editing (Cheng and Casey, unpublished data). Rather, HDV genotype I may limit the extent of editing via the inhibitory effect of HDAg-L on RNA replication, which is required for editing to occur on the antigenome and for the synthesis of edited genomic copies (Sato and Lazinski, personal communication).

Assays for Editing

Analysis of RNA editing at the HDV amber/W site has relied principally on two methods: comparison of HDAg-S and HDAg-L levels, and restriction digestion of PCR-amplified cDNA derived from HDV RNA. The former method has the advantage of being quick and simple and can be readily applied to analysis of editing on HDV RNA in cultured cells. However, this method is limited to cell-based experiments and by the requirements for a translated mRNA. The latter method works because editing at the amber/W site fortuitously creates a restriction digestion site that is not present in unedited RNA; enzymes used have included *Not* I, *Sty* I, *Dsa* I, *Btg* I. This method has the advantage of being more direct and can be applied to experiments performed in cells and in vitro. However, it is important to note that analysis of PCR products is susceptible to a potential artifact that could lead to an underestimate of editing levels. Because editing levels are frequently 30% or less, PCR products will be heterogeneous. If reannealing of these heterogeneous PCR products competes with primer annealing, then some PCR products will contain heteroduplexes in which one strand is derived from unedited RNA and the other from edited RNA. Such heteroduplexes will not be digested by the restriction enzyme and will result in underestimates of editing. To avoid this pitfall it is necessary to exclude heteroduplexes from the analysis.[22,55] One approach is to radioactively label PCR products only during the final extension step; in this way heteroduplexes are not labeled and do not contribute to the quantitation of products that are digested by the restriction enzyme. The accuracy of this approach has been verified by sequence analysis of cloned PCR products.[23,32,51]

Future Directions

Analysis of editing in HDV has led to valuable contributions to the field of RNA adenosine deamination. Thus far, it is the only example of specific editing that occurs in an organ other than the brain in mammals, but it is highly likely that more examples will be identified. While the general structures required for editing in genotypes I and III have been identified, the contributions of many elements - such as the numerous bulges on the 3' side of the amber/W site - in these structures to editing levels and specificity have yet to be fully explored. It seems likely that the editing site in genotype II will use the unbranched rod structure, as in genotype I, but this remains to be demonstrated. There is also much to learn about the compatibility of structures required for editing with those required for replication. Finally, given that both the structures required for editing and the C-terminal region of HDAg-L are defining features of HDV genotypes I, II, and III, it will be interesting to explore the relationship between editing and HDAg-L function.

Acknowledgements

I would like to thank Dr. Qiufang Cheng for comments on the manuscript. The work in the author's laboratory is supported by NIH grant AI42324.

References

1. Benne R, Van den Burg J, Brakenhoff JP et al. Major transcript of the frameshifted coxII gene from trypanosome mitochondria contains four nucleotides that are not encoded in the DNA. Cell 1986; 46(6):819-826.
2. Scott J. Messenger RNA editing and modification. Curr Opin Cell Biol Dec 1989; 1(6):1141-1147.
3. Higuchi M, Single FN, Kohler M et al. RNA editing of AMPA receptor subunit GluR-B: A base-paired intron-exon structure determines position and efficiency. Cell 1993; 75(7):1361-1370.
4. Curran J, Kolakofsky D. Sendai virus P gene produces multiple proteins from overlapping open reading frames. Enzyme 1990; 44(1-4):244-249.
5. Bonino F, Hoyer B, Ford E et al. The delta agent: HBsAg particles with delta antigen and RNA in the serum of an HBV carrier. Hepatology 1981; 1(2):127-131.
6. Bonino F, Hoyer B, Shih JW et al. Delta hepatitis agent: Structural and antigenic properties of the delta- associated particle. Infect Immun 1984; 43(3):1000-1005.
7. Bergmann KF, Gerin JL. Antigens of hepatitis delta virus in the liver and serum of humans and animals. J Infect Dis 1986; 154(4):702-706.
8. Bonino F, Heermann KH, Rizzetto M et al. Hepatitis delta virus: Protein composition of delta antigen and its hepatitis B virus-derived envelope. J Virol 1986; 58(3):945-950.
9. Wang KS, Choo QL, Weiner AJ et al. Structure, sequence and expression of the hepatitis delta viral genome. Nature 1986; 323(6088):508-514.
10. Makino S, Chang MF, Shieh CK et al. Molecular cloning and sequencing of a human hepatitis delta virus RNA. Nature 1987; 329(6137):343-346.
11. Kuo MY, Chao M, Taylor J. Initiation of replication of the human hepatitis delta virus genome from cloned DNA: Role of delta antigen. J Virol 1989; 63(5):1945-1950.
12. Chao M, Hsieh SY, Taylor J. Role of two forms of hepatitis delta virus antigen: Evidence for a mechanism of self-limiting genome replication. J Virol 1990; 64(10):5066-5069.
13. Chang FL, Chen PJ, Tu SJ et al. The large form of hepatitis delta antigen is crucial for assembly of hepatitis delta virus. Proc Natl Acad Sci USA 1991; 88(19):8490-8494.
14. Hwang SB, Lee CZ, Lai MM. Hepatitis delta antigen expressed by recombinant baculoviruses: Comparison of biochemical properties and post-translational modifications between the large and small forms. Virology 1992; 190(1):413-422.
15. Glenn JS, Watson JA, Havel CM et al. Identification of a prenylation site in delta virus large antigen. Science 1992; 256(5061):1331-1333.
16. Xia YP, Chang MF, Wei D et al. Heterogeneity of hepatitis delta antigen. Virology 1990; 178(1):331-336.

17. Weiner AJ, Choo QL, Wang KS et al. A single antigenomic open reading frame of the hepatitis delta virus encodes the epitope(s) of both hepatitis delta antigen polypeptides p24 delta and p27 delta. J Virol 1988; 62(2):594-599.

18. Sureau C, Taylor J, Chao M et al. Cloned hepatitis delta virus cDNA is infectious in the chimpanzee. J Virol 1989; 63(10):4292-4297.

19. Luo GX, Chao M, Hsieh SY et al. A specific base transition occurs on replicating hepatitis delta virus RNA. J Virol 1990; 64(3):1021-1027.

20. Casey JL, Bergmann KF, Brown TL et al. Structural requirements for RNA editing in hepatitis delta virus: Evidence for a uridine-to-cytidine editing mechanism. Proc Natl Acad Sci USA 1992; 89(15):7149-7153.

21. Zheng H, Fu TB, Lazinski D et al. Editing on the genomic RNA of human hepatitis delta virus. J Virol 1992; 66(8):4693-4697.

22. Casey JL, Gerin JL. Hepatitis D virus RNA editing: Specific modification of adenosine in the antigenomic RNA. J Virol 1995; 69(12):7593-7600.

23. Polson AG, Bass BL, Casey JL. RNA editing of hepatitis delta virus antigenome by dsRNA-adenosine deaminase. Nature 1996; 380(6573):454-456.

24. Yang JH, Sklar P, Axel R et al. Purification and characterization of a human RNA adenosine deaminase for glutamate receptor B premRNA editing. Proc Natl Acad Sci USA 1997; 94(9):4354-4359.

25. Melcher T, Maas S, Herb A et al. A mammalian RNA editing enzyme. Nature 1996; 379(6564):460-464.

26. O'Connell MA, Krause S, Higuchi M et al. Cloning of cDNAs encoding mammalian double-stranded RNA-specific adenosine deaminase. Mol Cell Biol 1995; 15(3):1389-1397.

27. Patterson JB, Samuel CE. Expression and regulation by interferon of a double-stranded-RNA-specific adenosine deaminase from human cells: Evidence for two forms of the deaminase. Mol Cell Biol 1995; 15(10):5376-5388.

28. Brusa R, Zimmermann F, Koh DS et al. Early-onset epilepsy and postnatal lethality associated with an editing- deficient GluR-B allele in mice. Science 1995; 270(5242):1677-1680.

29. Wang Q, Khillan J, Gadue P et al. Requirement of the RNA editing deaminase ADAR1 gene for embryonic erythropoiesis. Science 2000; 290(5497):1765-1768.

30. Seeburg PH. A-to-I editing: New and old sites, functions and speculations. Neuron 2002; 35(1):17-20.

31. Bass BL. RNA editing by adenosine deaminases that act on RNA. Annu Rev Biochem 2002; 71:817-846.

32. Jayan GC, Casey JL. Increased RNA editing and inhibition of hepatitis delta virus replication by high-level expression of ADAR1 and ADAR2. J Virol 2002; 76(8):3819-3827.

33. Wong SK, Sato S, Lazinski DW. Substrate recognition by ADAR1 and ADAR2. Rna 2001; 7(6):846-858.

34. Sato S, Wong SK, Lazinski DW. Hepatitis delta virus minimal substrates competent for editing by ADAR1 and ADAR2. J Virol 2001; 75(18):8547-8555.

35. Wong SK, Lazinski DW. Replicating hepatitis delta virus RNA is edited in the nucleus by the small form of ADAR1. Proc Natl Acad Sci USA 2002; 99(23):15118-15123.

36. Jayan GC, Casey JL. Inhibition of hepatitis delta virus RNA editing by short inhibitory RNA-mediated knockdown of Adar1 but not Adar2 expression. J Virol 2002; 76(23):12399-404.

37. Lehmann KA, Bass BL. The importance of internal loops within RNA substrates of ADAR1. J Mol Biol 1999; 291(1):1-13.

38. Polson AG, Bass BL. Preferential selection of adenosines for modification by double- stranded RNA adenosine deaminase. Embo J 1994; 13(23):5701-5711.

39. Ohman M, Kallman AM, Bass BL. In vitro analysis of the binding of ADAR2 to the premRNA encoding the GluR-B R/G site. Rna 2000; 6(5):687-697.

40. Herbert A, Rich A. The role of binding domains for dsRNA and Z-DNA in the in vivo editing of minimal substrates by ADAR1. Proc Natl Acad Sci USA 2001; 98(21):12132-12137.

41. Lomeli H, Mosbacher J, Melcher T et al. Control of kinetic properties of AMPA receptor channels by nuclear RNA editing. Science 1994; 266(5191):1709-1713.

42. Herb A, Higuchi M, Sprengel R et al. Q/R site editing in kainate receptor GluR5 and GluR6 premRNAs requires distant intronic sequences. Proc Natl Acad Sci USA 1996; 93(5):1875-1880.
43. Shakil AO, Hadziyannis S, Hoofnagle JH et al. Geographic distribution and genetic variability of hepatitis delta virus genotype I. Virology 1997; 234(1):160-167.
44. Niro GA, Smedile A, Andriulli A et al. The predominance of hepatitis delta virus genotype I among chronically infected Italian patients. Hepatology 1997; 25(3):728-734.
45. Casey JL. RNA editing in hepatitis delta virus genotype III requires a branched double-hairpin RNA structure. J Virol 2002; 76(15):7385-7397.
46. Hsu SC, Syu WJ, Sheen IJ et al. Varied assembly and RNA editing efficiencies between genotypes I and II hepatitis D virus and their implications. Hepatology 2002; 35(3):665-672.
47. Casey JL, Brown TL, Colan EJ et al. A genotype of hepatitis D virus that occurs in northern South America. Proc Natl Acad Sci USA 1993; 90(19):9016-9020.
48. Ivaniushina V, Radjef N, Alexeeva M et al. Hepatitis delta virus genotypes I and II cocirculate in an endemic area of Yakutia, Russia. J Gen Virol 2001; 82(Pt 11):2709-2718.
49. Yang A, Papaioannou C, Hadzyannis S et al. Base changes at positions 1014 and 578 of delta virus RNA in Greek isolates maintain base pair in rod conformation with efficient RNA editing. J Med Virol 1995; 47(2):113-119.
50. Jayan GC, Casey JL. Unpublished.
51. Polson AG, Ley 3rd HL, Bass BL et al. Hepatitis delta virus RNA editing is highly specific for the amber/W site and is suppressed by hepatitis delta antigen. Mol Cell Biol 1998; 18(4):1919-1926.
52. Aruscavage PJ, Bass BL. A phylogenetic analysis reveals an unusual sequence conservation within introns involved in RNA editing. RNA 2000; 6(2):257-269.
53. Glenn JS, White JM. trans-dominant inhibition of human hepatitis delta virus genome replication. J Virol 1991; 65(5):2357-2361.
54. Cheng Q, Jayan GC, Casey JL. Differential inhibition of RNA editing in hepatitis delta virus genotype III by the short and long forms of hepatitis delta antigen. J Virol Jul 2003; 77(14):7786-7795.
55. Wu TT, Bichko VV, Ryu WS et al. Hepatitis delta virus mutant: Effect on RNA editing. J Virol 1995; 69(11):7226-7231.
56. Zuker M, Mathews DH, Turner DH. Algorithms and thermodynamics for RNA secondary structure prediction: A practical guide. In: Barciszewski J, Clark BFC, eds. RNA Biochemistry and Biotechnology. Kluwer Academic Publishers, 1999:11-43.
57. Seeburg PH, Higuchi M, Sprengel R. RNA editing of brain glutamate receptor channels: Mechanism and physiology. Brain Res Brain Res Rev 1998; 26(2-3):217-229.

CHAPTER 6

Hepatitis Delta Antigen and RNA Polymerase II

Yuki Yamaguchi and Hiroshi Handa*

Abstract

R eplication and transcription of HDV proceed via RNA-dependent RNA synthesis. These reactions are thought to be catalyzed at least in part by host RNA polymerase II (RNAPII). Hepatitis delta antigen (HDAg), which is critical for these processes, was recently proposed to function as a transcription elongation factor for RNAPII. The involvement of a DNA-dependent RNA polymerase in RNA-dependent RNA synthesis is itself intriguing and poses fundamental questions as to how RNA synthesis initiates, elongates, and terminates on an unusual HDV RNA template. In addition, the presence of a 'viral' transcription elongation factor is unprecedented in eukaryotes, whereas a few are known to exist in prokaryotes. Thus, the study of HDV replication and transcription should provide tremendous insight into the basic mechanism underlying RNAPII transcription.

Introduction

Three types of RNA-dependent RNA synthesis occur during the HDV life cycle: (i) antigenomic RNA synthesis from genomic RNA, (ii) genomic RNA synthesis from antigenomic RNA, and (iii) HDAg mRNA synthesis from genomic RNA (see Chapter 3 for details). The first and second types of reactions are steps in replication that are thought to proceed by a 'rolling cycle' mechanism. This mechanism is analogous to DNA replication of many plasmids and filamentous bacteriophages. As for the third type of reaction, based on the analysis of the mRNA's 5' end, it is assumed that the transcription is initiated from a position that is very close to an end of the rod-like structure of the HDV genome. By extension, the first type of reaction, which also utilizes genomic RNA as a template, may be initiated from the same position of the HDV genome. In this chapter, we refer to the three types of reactions simply as 'transcription'.

Several lines of evidence suggest that RNAPII is involved in HDV RNA transcription. First, viroid RNAs, infectious agents in plants that show structural similarity to HDV RNA, are reportedly transcribed by RNAPII in cell-free extracts.[1] Second, as reported by a few laboratories, HDV RNA can also be transcribed by RNAPII in vitro.[2-4] It should be noted, however, that the studies completed thus far have been unable to synthesize full-length complementary RNAs. In one report, for example, RNAPII in the nuclear extract of human HeLa cells directed a genomic strand synthesis of up to ~40 nt using an antigenomic fragment of HDV RNA as a

*Corresponding Author: Hiroshi Handa—Graduate School of Bioscience and Biotechnology, Tokyo Institute of Technology, 4259 Nagatsuta, Yokohama 226-8501, Japan. Email: hhanda@bio.titech.ac.jp

Hepatitis Delta Virus, edited by Hiroshi Handa and Yuki Yamaguchi.
©2006 Landes Bioscience and Springer Science+Business Media.

template.[3] Third, HDAg, the sole HDV protein, directly binds to RNAPII and remarkably stimulates DNA-directed and HDV RNA-directed transcription in vitro.[5-7] The second half of this chapter deals with this topic. Fourth, HDAg mRNA is capped and polyadenylated at its 5' and 3' ends, respectively.[4] These processing events are tightly coupled to RNAPII transcription and occur in all the known mRNA species synthesized by RNAPII, with the exception of histone mRNA. Conversely, essentially no RNA species synthesized by other RNA polymerases are capped or polyadenylated. Fifth, in intact cells and in isolated nuclei, transcription of the HDV genome is reportedly sensitive to the mushroom toxin α-amanitin at concentrations low enough to selectively inhibit RNAPII.[8,9] One may need to view this with caution, however, because opposing results have been presented by another laboratory[10,11] (see Chapter 3 for more discussion). With these findings taken together, it should be reasonable to conclude that RNAPII is responsible at least in part for HDV RNA transcription.

Variation on a Theme: Initiation, Elongation, and Termination of HDV RNA Transcription

The idea that RNAPII, a DNA-dependent RNA polymerase, directs RNA-dependent RNA synthesis poses several interesting questions as to how RNA synthesis initiates, elongates, and terminates on an unusual HDV RNA template. From a mechanistic point of view, such an RNA-directed transcription seems quite a challenge to RNAPII, as discussed below. Elucidation of this mechanism may lead to the identification of new molecular targets to prevent the pathogenic virus. Furthermore, such knowledge should add insightful information on the basic mechanism of RNAPII transcription.

Before moving on to the central issue, we first overview the process of DNA-directed transcription by RNAPII. The transcription process comprises several distinct steps, including: (i) preinitiation complex assembly, (ii) promoter opening, (iii) transcription initiation, (iv) promoter escape, (v) transcription elongation, and (vi) transcription termination (Fig. 1A). The first four steps occur around transcription initiation sites. RNAPII alone is unable to initiate transcription. Instead it forms a preinitiation complex together with general transcription factors, including transcription factor (TF) IIA, TFIIB, TFIID, TFIIE, TFIIF, and TFIIH, on a promoter.[12] Core promoter elements, such as TATA boxes and Inr elements, are important for the assembly. TFIIH then facilitates the conversion of a closed-to-open complex by its DNA helicase activity in an ATP-dependent manner (promoter opening).[12] Next, RNAPII starts to synthesize nascent RNA but immediately encounters a transcriptional block when it reaches 9~12 bp downstream. TFIIH helicase suppresses the block and facilitates the transition to the elongation phase (promoter escape).[13] This step is equated with the dissociation of RNAPII from promoter-bound transcription factors. During transcription elongation, RNAPII forms a ternary complex together with template DNA and nascent RNA. Within the 'transcription elongation complex', 12-15 bp of DNA are unwound to form a 'transcription bubble'. In addition, 8-9 nt of RNA in the 3'-end are contained by forming a hybrid with the template stand of DNA, with the growing 3'-end usually maintained at the active site of RNAPII.[14] Termination of premRNA synthesis is tightly but not entirely coupled to 3'-end processing.[15] The processing event, composed of transcript cleavage and polyadenylation, takes place 23 or 24 nt downstream of the AAUAAA sequence. Transcription termination seems to occur rather randomly between 200-2000 bp downstream of the poly(A) signal, triggered by preceding transcript cleavage and polyadenylation.

How is transcription initiated on the HDV RNA template? A few laboratories investigated the requirement for cis-acting elements in HDV transcription in vitro. According to these studies, small bulges close to the ends of the rod-like genome, where transcription is considered to be initiated, are important for efficient transcription.[2,3] These studies, however, were unable to establish the necessary and sufficient conditions for transcription initiation in

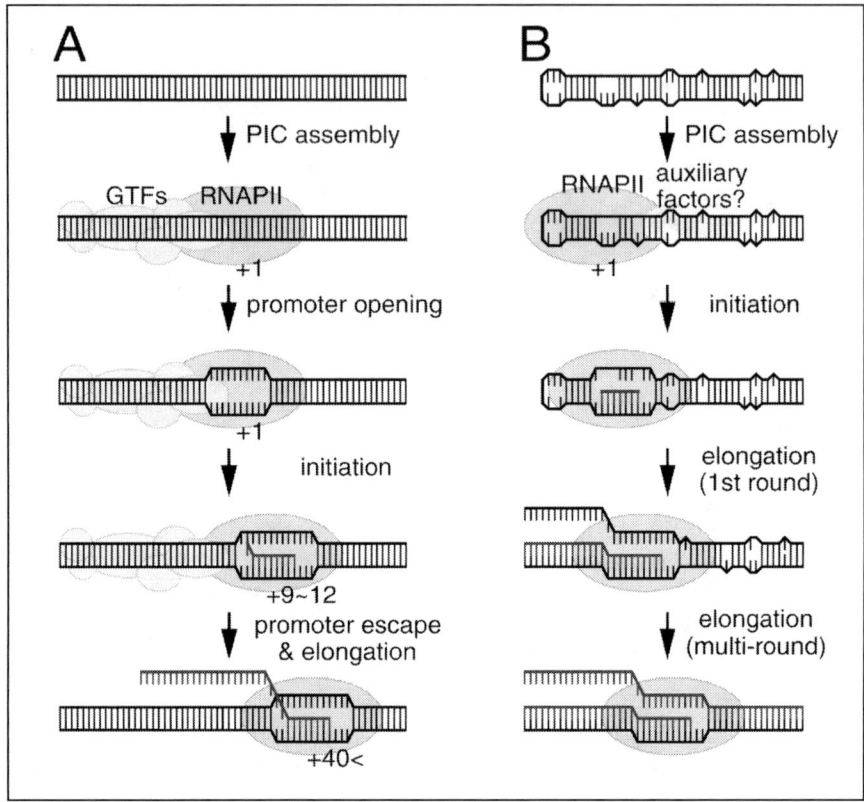

Figure 1. Comparison of DNA-directed (A) and HDV RNA-directed (B) transcription by RNAPII. The termination step is omitted in the illustration. DNA/RNA templates and nascent RNA transcripts are drawn in black and blue, respectively. Numerals indicate positions of the RNAPII active site relative to the transcription initiation site. GTF, general transcription factor; PIC, preinitiation complex.

sufficient detail. Moreover, virtually nothing is known about trans-acting factors involved in the initiation process. One can postulate several possibilities. First, RNAPII recruitment and transcription initiation may require some or all of the general transcription factors, and both processes proceed in manners analogous to those of DNA-directed transcription. Of the general transcription factors, at least TFIID and TFIIB bind to DNA with sequence specificity. It is not known whether they can also bind to RNA. Second, protein factors other than the general transcription factors, such as cellular RNA-binding proteins and the viral HDAg, may be involved in the initiation process (Fig. 1B). Third, RNAPII alone may be sufficient to initiate transcription on an unusual template.

Perhaps, data obtained using special DNA templates may provide clues to the solution. On a DNA template containing a bubble or nonbase-paired region of ~10 nt at a promoter, RNAPII initiates transcription without some of the general transcription factors, namely TFIIE and TFIIH[16] (Fig. 2). On the other hand, on a linear DNA template carrying a 3'-overhang of oligo(dC), RNAPII alone can initiate transcription from its end efficiently[17] (Fig. 2). Apparently, these DNA templates show some degree of structural similarity to the HDV genome. Even if general transcription factors are involved in the HDV RNA-directed transcription, we suspect that the factor requirement may be quite different from that of DNA-directed transcription.

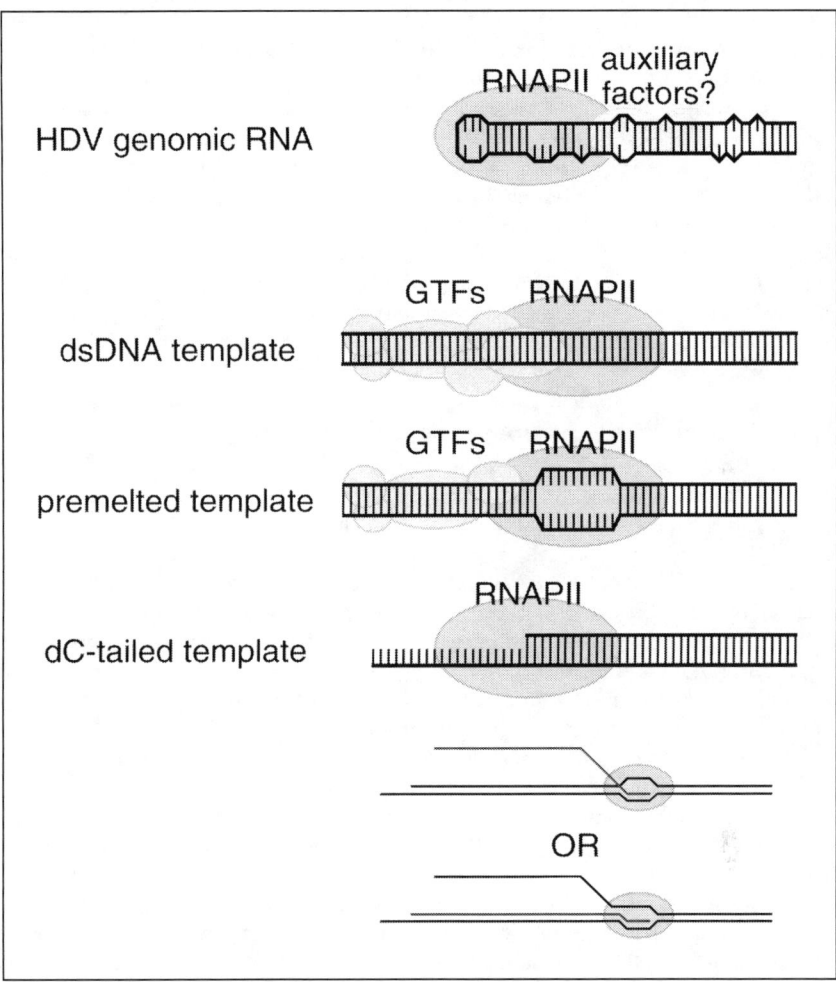

Figure 2. Comparison of the rod-like HDV genome and DNA templates used in in vitro study. Normal double-stranded (ds) DNA template, premelted template, and dC-tailed template show different factor requirements in transcription initiation. Also shown at the bottom are two modes of transcription elongation on a dC-tailed template. See text for detail. DNA/RNA templates and nascent RNA transcripts are drawn in black and blue, respectively. GTF, general transcription factor.

How does a nascent RNA chain elongate on the HDV RNA template? As transcription proceeds by the rolling circle mechanism, a nascent transcript forms a long double-stranded region with the RNA template (Fig. 1B). By comparison, during accurately initiated transcription on a DNA template, a nascent transcript is separated from DNA within the transcription elongation complex, and only 8-9 bases of RNA form a hybrid with the template strand of DNA (Fig. 1A). RNAPII seems to be intrinsically capable of both types of RNA synthesis. During transcription of a dC-tailed template, sometimes a significant fraction of the nascent transcripts displace the nontemplate strand of DNA to form long DNA-RNA hybrids[17] (Fig. 2). The efficiency of this unusual synthesis pathway seems affected in part by the G/C content of the initially transcribed sequence.[17]

Another consideration concerns structural differences between DNA and RNA, which may pose a considerable problem for the progression of RNAPII. HDV RNA's differences from DNA include 2'-OH versus 2'-H, double helices in an A form versus a B form, and the presence versus absence of bulges and single-stranded regions that interrupt double helices. Data obtained with DNA templates have shown that a small change in the nucleic acid structure often causes a strong transcriptional block.[18,19] Thus, positive transcription elongation factors that suppress the transcriptional block and enhance the efficiency of elongation may play important roles in HDV RNA-directed transcription, as discussed later.

How does transcription terminate on the HDV RNA template? In the case of the synthesis of HDAg mRNA, it is very likely that the termination process is coupled with 3'-end processing. As evidence, the HDV genome encodes canonical poly(A) signals, and the viral mRNA is polyadenylated.[4] By contrast, little is known as to where and how the syntheses of the multimeric forms of the HDV genome and antigenome terminate.

HDAg as a Viral Transcription Elongation Factor

RNAPII elongation on a DNA template is controlled by over a dozen transcription elongation factors.[20] Strictly defined, a transcription elongation factor is a protein that directly associates with RNA polymerase and thereby regulates its elongation activity.

A class of transcription elongation factors, including TFIIF, TFIIS, Elongin, and ELL, interacts with RNAPII and accelerates elongation (TFIIF plays dual roles in transcription initiation and elongation) (Fig. 3A). This class of factors probably does so in part by preventing RNAPII from frequent pauses and arrests caused by inhibitory structures of nascent RNA, low concentrations of NTP, and misincorporation of NMP. In this context, a pause is a transient inactive state, whereas an arrest is a more persistent state that eventually leads to termination without reactivation by positive transcription elongation factors.

Another class of transcription elongation factors includes DRB sensitivity-inducing factor (DSIF), negative elongation factor (NELF), and positive transcription elongation factor b (P-TEFb), which together constitute a critical rate-limiting step of transcription[20,21] (Fig. 3B). Shortly after the initiation of transcription, RNAPII is subjected to both negative and positive control by these factors. DSIF and NELF together cause a transcriptional block by binding to RNAPII. Conversely, P-TEFb allows RNAPII to enter a productive elongation phase by preventing DSIF and NELF from acting. P-TEFb is a protein kinase that strongly phosphorylates the C-terminal domain of RNAPII and the Spt5 subunit of DSIF[22] (Fig. 4A). Several, but not all, lines of evidence suggest that P-TEFb-dependent phosphorylation of the CTD facilitates the release of NELF from RNAPII, thereby overcoming the transcriptional block[21,23] (Fig. 4A).

Recent papers have presented biochemical evidence for the role of HDAg in RNAPII elongation.[5-7] NELF is composed of five polypeptides, and one of these subunits, called NELF-A, was found to show limited sequence similarity to HDAg.[5] This similarity stimulated subsequent study and led to the following findings. First, HDAg directly binds to RNAPII in a manner competitive to NELF[5] (Fig. 4B). The regions of similarity shared between HDAg and NELF-A may form a conserved structure that recognizes a common surface on RNAPII. This idea is supported by the results of a more recent deletion analysis of NELF-A.[24] Second, HDAg activates RNAPII elongation on an HDV RNA template as well as on a DNA template in vitro[5-7] (Fig. 4B). When a dC-tailed template and pure RNAPII are used, for example, HDAg remarkably enhances the rate of elongation regardless of the presence of DSIF and NELF. Thus, the activation mechanism appears two-fold: HDAg reverses the negative effect of DSIF and NELF by displacing NELF from RNAPII, and the HDAg-RNAPII interaction by itself further activates transcription elongation. HDAg can be judged on several criteria as a new member, and in fact it is the first viral member of the transcription elongation factors for RNAPII.

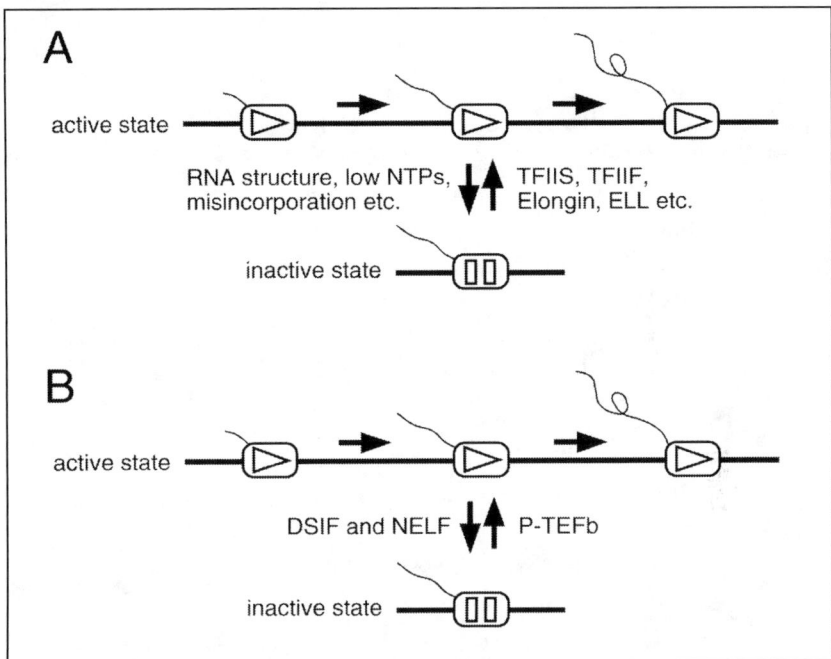

Figure 3. Two modes of elongation control. A) A class of transcription elongation factors, including TFIIF, TFIIS, Elongin, and ELL, interacts with RNAPII and accelerates elongation by preventing RNAPII from frequent pauses and arrests caused by inhibitory structures of nascent RNA, low concentrations of NTP, and misincorporation of NMP. B) Another class of transcription elongation factors, including DSIF, NELF, and P-TEFb, constitutes a critical rate-limiting step of transcription. See text for detail.

HDAg is thought to be a multifunctional protein playing important roles in nucleocyto-plasmic transport of HDV RNA, transcriptional regulation, and viral assembly, among others (see Chapters 3 and 4). Therefore, dissection of the protein motifs responsible for each function is important in order to understand the relative significance of these functions in the cellular context. Deletion and mutagenic analyses of HDAg have found the following: Two-thirds of HDAg protein, containing the oligomerization domain, the nuclear localization signal, and the Arg-rich RNA-binding motif, are dispensable for RNAPII-binding and elongation stimulation activities in vitro (5 and our unpublished data). On the other hand, the C-terminal region of HDAg, which is well conserved among different HDV genotypes and among clinical isolates of HDV, is involved in HDAg's interaction with RNAPII.[5] Point mutations of the conserved amino acid residues within this region strongly reduce the affinity of the HDAg-RNAPII interaction in vitro (our unpublished data). The same set of mutations also impairs HDAg's ability to support HDV replication in cell culture (our unpublished data), suggesting that RNAPII binding and activation by HDAg is important for HDV replication in vivo.

Based on the available data, we present a hypothetical model as to how HDAg activates HDV RNA transcription. In vitro transcription systems that use a genomic fragment of HDV RNA and HeLa cell nuclear extracts have been often criticized for their low efficiency and their inability to synthesize full-length complementary RNAs.[4] These defects are most likely attributable to the lack of HDAg. In one report, RNAPII stopped transcription after synthesizing a ~40 nt complementary RNA in the absence of HDAg.[5] On a DNA template, DSIF and NELF are known to repress an early step of elongation, primarily around 30~50 nt downstream of the

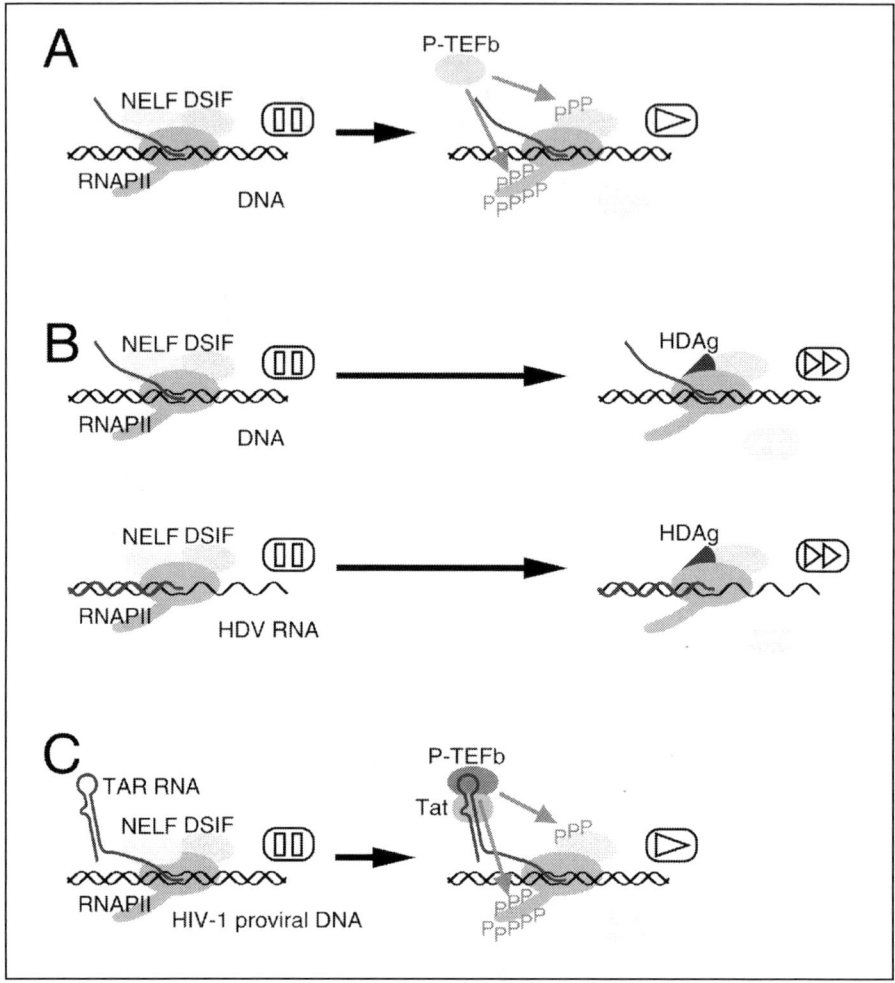

Figure 4. HDAg is a viral transcription elongation factor. A) DSIF and NELF bind to RNAPII and repress its elongation. P-TEFb phosphorylates RNAPII and DSIF to reverse the repression. B) According to a proposal,[5] HDAg reverses the negative effect of DSIF and NELF by displacing NELF from RNAPII, and the HDAg-RNAPII interaction by itself further activates transcription elongation. C) HIV Tat binds to TAR on an HIV transcript, recruits P-TEFb, and allows efficient reversal of the DSIF and NELF repression. DNA/RNA templates and nascent RNA transcripts are drawn in black and blue, respectively.

transcription initiation site.[21] The close correspondence of the stalled positions suggests that HDAg stimulates HDV RNA transcription by overcoming the repression imposed by DSIF and NELF. At further downstream sites, HDAg may also stimulate elongation by suppressing pauses or arrests that are induced by unfavorable structures of the RNA template or transcript.

As for the two isoforms of HDAg, it is generally thought that HDAg-S activates HDV transcription, whereas HDAg-L represses HDV transcription, possibly through its dominant-negative effect on HDAg-S, and directs viral assembly. In vitro transcription assays using bacterially expressed recombinant HDAg proteins have shown that HDAg-L is capable of activating transcription, even though it is several times less active than HDAg-S.[5] Naturally, that experiment does not consider post-translational modifications such as phosphorylation and prenylation. Especially, prenylation is known to occur within the L-specific peptide.[25] Such modifications in vivo might make HDAg-L transcriptionally inactive either directly, by preventing its interaction with RNAPII, or indirectly, by sequestering it to the cytoplasm, and thereby allow HDAg-L to inhibit HDV transcription in a dominant-negative manner.

Molecular Analysis of Elongation Control by HDAg

The small size and robust effect of HDAg make it an ideal model for understanding the molecular mechanism underlying elongation control. In recent papers, the elongation process on immobilized DNA templates has been analyzed in detail at nucleotide resolutions of milli-second time resolution.[6,7] At least four chemical steps are required for each nucleotide addition: NTP binding to the RNAPII active site, forward translocation of RNAPII, phosphodiester bond formation, and pyrophosphate release. Kinetic analysis shows that there is a rate-limiting step during each nucleotide addition, and that this step is activated by HDAg. Circumstantial evidence suggests that this step is equal to translocation. HDAg may thus facilitate the otherwise slow translocation step to accelerate RNAPII elongation.

The structural basis of the RNAPII-HDAg interaction is now being investigated (our unpublished data). RNAPII is a half megadalton, twelve-subunit protein complex. Since the crystal structure of yeast RNAPII was solved in 2001,[26,27] much attention has been focused on the interactions between RNAPII and accessory proteins. Given the situation that crystallization of human RNAPII is still a long way off, an alternative approach is site-specific photocrosslinking combined with proteolytic mapping. This method requires site-specific introduction of a photoactivatable crosslinker to a protein of interest, usually through a unique cysteine residue. Unlike the known 'cellular' transcription elongation factors, HDAg is a small protein consisting of only 195 amino acids and with no cysteine. Therefore, the crosslinker can be easily introduced at a desired position by cysteine substitution. Initial data suggest that the C-terminal region of HDAg contacts with the RNAPII Rpb1 and Rpb2 subunits, which consist of a 'clamp'. The clamp is a mobile structure that grips DNA and RNA during elongation.[28] An interesting hypothesis is that HDAg may bind and loosen the clamp not only to facilitate RNAPII translocation, as mentioned above, but also to influence template recognition. During HDV RNA transcription, RNAPII is converted from a DNA-dependent RNA polymerase to a dual-specificity RNA polymerase. This loss of template selectivity may be accounted for by HDAg loosening the clamp.

Targeting Transcription Elongation: A General Strategy for Viruses?

In the last section, we pointed out an interesting functional similarity between HDAg and other viral proteins in terms of elongation control. The Tat protein encoded by human immunodeficiency virus (HIV) is one of such examples. HIV establishes latent infection following provirus integration into the host genome. Tat plays a critical role in activating HIV transcription by host RNAPII and thereby regulates the switch from latent to productive infection. Extensive studies have shown that Tat binds to TAR, an RNA element located on the nascent transcript, and recruits P-TEFb to the promoter-proximal region. P-TEFb then facilitates the synthesis of full-length HIV transcripts, in part by counteracting the repression imposed by DSIF and NELF (Fig. 4C). Thus, HDAg and Tat exert their functions by targeting the same transcriptional regulatory machinery in different ways. The N protein encoded by bacteriophage λ is another example. N plays an important role in switching from the lysogenic to the lytic cycle through elongation control, better known as termination and antitermination in prokaryotes. N binds to nut, which are RNA elements located on nascent transcripts of λ operons. N also interacts with cellular transcription factors and RNA polymerase, prevents termination at multiple downstream sites, and allows the production of full-length transcripts encoding the 'late' genes. Thus, controlling transcription elongation may be viruses' general strategy for successfully completing their life cycle.

References

1. Fels A, Hu K, Riesner D. Transcription of potato spindle tuber viroid by RNA polymerase II starts predominantly at two specific sites. Nucleic Acids Res 2001; 29:4589-4597.
2. Beard MR, MacNaughton TB, Gowans EJ. Identification and characterization of a hepatitis delta virus RNA transcriptional promoter. J Virol 1996; 70:4986-4995.
3. Filipovska J, Konarska MM. Specific HDV RNA-templated transcription by pol II in vitro. RNA 2000; 6:41-54.
4. Gudima S, Wu SY, Chiang CM et al. Origin of hepatitis delta virus mRNA. J Virol 2000; 74:7204-7210.
5. Yamaguchi Y, Filipovska J, Yano K et al. Stimulation of RNA polymerase II elongation by hepatitis delta antigen. Science 2001; 293:124-127.
6. Nedialkov YA, Gong XQ, Hovde SL et al. NTP-driven translocation by human RNA polymerase II. J Biol Chem 2003; 278:18303-18312.
7. Zhang C, Yan H, Burton ZF. Combinatorial control of human RNA polymerase II pausing and transcript cleavage by Transcription Factor IIF, hepatitis delta antigen and TFIIS/SII. J Biol Chem 2003; 278(50):50101-11.
8. MacNaughton TB, Gowans EJ, McNamara SP et al. Hepatitis delta antigen is necessary for access of hepatitis delta virus RNA to the cell transcriptional machinery but is not part of the transcriptional complex. Virology 1991; 184:387-390.
9. Moraleda G, Taylor J. Host RNA polymerase requirements for transcription of the human hepatitis delta virus genome. J Virol 2001; 75:10161-10169.
10. Modahl LE, Macnaughton TB, Zhu N et al. RNA-Dependent replication and transcription of hepatitis delta virus RNA involve distinct cellular RNA polymerases. Mol Cell Biol 2000; 20:6030-6039.
11. Macnaughton TB, Shi ST, Modahl LE et al. Rolling circle replication of hepatitis delta virus RNA is carried out by two different cellular RNA polymerases. J Virol 2002; 76:3920-3927.
12. Fiedler U, Marc Timmers HT. Peeling by binding or twisting by cranking: Models for promoter opening and transcription initiation by RNA polymerase II. Bioessays 2000; 22:316-326.
13. Dvir A. Promoter escape by RNA polymerase II. Biochim Biophys Acta 2002; 1577:208-223.
14. Gnatt A. Elongation by RNA polymerase II: Structurefunction relationship. Biochim Biophys Acta 2002; 1577:175-190.
15. Proudfoot NJ, Furger A, Dye MJ. Integrating mRNA processing with transcription. Cell 2002; 108:501-12.
16. Pan G, Greenblatt J. Initiation of transcription by RNA polymerase II is limited by melting of the promoter DNA in the region immediately upstream of the initiation site. J Biol Chem 1994; 269:30101-30104.
17. Kadesch TR, Chamberlin MJ. Studies of in vitro transcription by calf thymus RNA polymerase II using a novel duplex DNA template. J Biol Chem 1982; 257:5286-5295.
18. Tornaletti S, Donahue BA, Reines D et al. Nucleotide sequence context effect of a cyclobutane pyrimidine dimer upon RNA polymerase II transcription. J Biol Chem 1997; 272:31719-31724.
19. Kuraoka I, Endou M, Yamaguchi Y et al. Effects of endogenous DNA base lesions on transcription elongation by mammalian RNA polymerase II. Implications for transcription-coupled DNA repair and transcriptional mutagenesis. J Biol Chem 2003; 278:7294-7299.
20. Conaway JW, Shilatifard A, Dvir A et al. Control of elongation by RNA polymerase II. Trends Biochem Sci 2000; 25:375-380.
21. Yamaguchi Y, Takagi T, Wada T et al. NELF, a multisubunit complex containing RD, cooperates with DSIF to repress RNA polymerase II elongation. Cell 1999; 97:41-51.
22. Kim JB, Sharp PA. Positive transcription elongation factor B phosphorylates hSPT5 and RNA polymerase II carboxyl-terminal domain independently of cyclin-dependent kinase-activating kinase. J Biol Chem 2001; 276:12317-12323.

23. Wu CH, Yamaguchi Y, Benjamin LR et al. NELF and DSIF cause promoter proximal pausing on the hsp70 promoter in Drosophila. Genes Dev 2003; 17:1402-1414.

24. Narita T, Yamaguchi Y, Yano K et al. Human transcription elongation factor NELF: Identification of novel subunits and reconstitution of the functionally active complex. Mol Cell Biol 2003; 23:1863-1873.

25. Glenn JS, Watson JA, Havel CM et al. Identification of a prenylation site in delta virus large antigen. Science 1992; 256:1331-1333.

26. Cramer P, Bushnell DA, Kornberg RD. Structural basis of transcription: RNA polymerase II at 2.8 angstrom resolution. Science 2001; 292:1863-1876.

27. Gnatt AL, Cramer P, Fu J et al. Structural basis of transcription: An RNA polymerase II elongation complex at 3.3 A resolution. Science 2001; 292:1876-1882.

CHAPTER 7

Clinical Features of Hepatitis Delta Virus

Dimitrios Vassilopoulos* and Stephanos J. Hadziyannis

Introduction

In the early days following the discovery of the hepatitis delta virus (HDV) much emphasis was given on the severity of delta hepatitis and its rapid progression to cirrhosis and liver failure. However with time going on, evidence started to accumulate indicating that in several individuals, chronic HDV infection could run a benign course, with silent clinical and even biochemical features and that in such patients liver histology would be more consistent with the mild changes of chronic persistent hepatitis rather than with the severe necroinflammation and advanced fibrosis of chronic active hepatitis.[1-4] In particular the search for serological markers of HDV infection in the general population of several communities worldwide and among blood donors (Japan, Taiwan, Greece and Italy)[3,4] revealed that the actual spectrum of delta hepatitis is very wide and heterogeneous and that similarly to the infection with the other known hepatitis viruses it can range from a very mild, clinically latent disease to florid active hepatitis and decompensated cirrhosis. With time going on and with the accumulation of new data from long-term follow-up studies[4,5] it also became obvious that the natural course of acute and chronic HDV infection is extremely variable and includes all possibilities from complete cure and burning out to slow progression, rapid progression, development of cirrhosis and liver failure and development of hepatocellular carcinoma (HCC). Furthermore with the application of refined serological, virological and other laboratory techniques, the clinical aspects of HDV infection could be associated and linked meaningfully with numerous viral, host and other variables.

In this article an attempt is made to describe the evolution over the years of our concepts on the clinical correlates and syndromes developing in acute and chronic HDV infection and on their natural course. In this context it is important to stress that hepatitis delta represents infection not with one but with two viruses (the HDV and the HBV) that are transmitted to the host either concomitantly (coinfection) or in the case of HDV superinfection in a host with preexisting chronic HBV infection. Moreover, due to epidemiological reasons and common risk factors coinfection or superinfection with the hepatitis C virus (HCV) and with the human immunodeficiency virus (HIV) is also encountered in clinical practice. Complex clinical features may thus arise while interactions between the coinfecting viruses may have significant impact on their replicative activity, on immune and pathogenetic mechanisms and consequently on the severity, course and outcome of the nosological syndromes resulting from each of them.

*Corresponding Author: Dimitrios Vassilopoulos—Athens University School of Medicine, Hippokration General Hospital, Academic Department of Medicine, 114 Vass. Sophias Ave., 115 27 Athens, Greece. Email: dvassilop@med.uoa.gr

Hepatitis Delta Virus, edited by Hiroshi Handa and Yuki Yamaguchi.
©2006 Landes Bioscience and Springer Science+Business Media.

Acute Hepatitis Delta

Infection with the hepatitis D virus of susceptible individuals results in acute hepatitis provided that the HDV gets the necessary helper function from the surface proteins of the HBV particularly the preS1 in order to achieve entry into hepatocytes and replicate. This condition is met either (a) by concomitant transmission of the two viruses to susceptible individuals lacking anti-HBV immunity (coinfection) or (b) by transmission of HDV to individuals with preexisting chronic HBV infection (HDV superinfection). Individuals with natural or vaccination-acquired immunity against HBV, harboring antibodies to the hepatitis B surface protein (anti-HBs), are protected from HDV infection.

Depending on the type of HDV infection (superinfection or coinfection) and on several host and viral factors like age, sex, immune status of the host, size of the viral inoculum, HBeAg/anti-HBe status of the infecting and infected individuals (in case of superinfection) and probably on the HDV and HBV genotypes, the acute HBV infection may attain mild, severe or even fulminant course, may resolve or may progress to chronicity.

Acute Hepatitis Due to Coinfection with HDV and HBV

The clinical features of acute hepatitis D due to coinfection are not much different from those of HDV superinfection. However in both conditions severe and even fulminant forms of hepatitis appear to occur more frequently compared with ordinary acute hepatitis B.[4,6] On the other hand, acute hepatitis D is indistinguishable from ordinary hepatitis B on clinical and histological grounds. The differential diagnosis is possible only on the basis of serological assays showing the presence of markers of HDV infection together with markers of primary HBV infection (high titers of IgM anti-HBc). The clinical expression may be biphasic with two episodes of acute hepatitis occurring a few weeks apart. The first is linked to peak HBV replication and the second to HDV replication. The HDV may suppress the replication of HBV and milder forms of acute hepatitis may, thus, occur in coinfection. Early suppression of HBV has been observed in Italy to inhibit the synthesis of HBsAg resulting in non detectability of this marker in serum.[7] In a recent study, a transient decrease in the levels of HBV DNA was observed in HBV/HDV acutely coinfected patients compared to patients with acute hepatitis B.[8]

Coinfections with HDV and HBV contrary to HDV superinfections of chronic HBsAg carriers also are known to be self limited in their vast majority.[9] This difference is attributed to the duration of infection with the helper hepatitis B virus which, being transient in coinfection, does not permit the HDV to outlive the episode of acute hepatitis. Why HBV infection is transient in coinfection can be explained better now by the fact that most coinfections have been identified among adults in whom, unlike the pattern observed in children, acute hepatitis B is self limited in more than 95% of its cases.[6] Moreover in most cases of coinfection the transmitted HBV strain is HBeAg-negative (usually precore mutant) and, as known, acute infection with HBeAg-negative HBV strains may cause severe acute and even fulminant hepatitis but rarely if ever chronic hepatitis B, irrespective of the age of patients.[3] In communities with endemic HBV infection self-limited coinfections with HBV and HDV appear to be quite frequent. A significant number of individuals who are positive for serological markers of past HBV infection (anti-HBc and anti-HBs positive) are also positive for serum anti-HDV of IgG class, a marker of past HDV infection.[10]

HDV RNA can be detected in serum in 93% to 100% of coinfected individuals.[11] At the same time, HBV is expressed in a lower percentage of patients and becomes undetectable with time. With resolution of HBV/HDV coinfection HDV RNA becomes undetectable, serum HBsAg is cleared and markers of past HBV and HDV infection develop.

Acute Hepatitis Due to HDV Superinfection

In this setting chronic HBV infection already preexists with variable effects on liver function and structure ranging from the inactive HBsAg carrier state to chronic hepatitis B with necroinflammation and advanced fibrosis including cirrhosis. The underlying liver condition is assumed to determine to a large extent the severity and outcome of HDV superinfection. In general superinfection usually results in overt acute hepatitis and if the preexisting HBV-induced liver damage is already advanced it may lead rapidly to liver failure. Cases of fulminant hepatitis D have been reported in the tropical Amazon Area of South America, due to superinfection with HDV genotype III of HBV carriers with genotype F.[12] Similar cases of fulminant hepatitis D superinfection (genotype I) have been recently reported in Russian patients.[13]

HDV superinfection usually becomes chronic (> 90%) because the underlying HBV infection is long-lived and continues to offer to HDV helper function for ongoing replication.[4] The concept that transition of superinfection to chronicity is mainly a function of HBV finds further support from the data on HBV DNA and HDV RNA in studies of the natural course of HDV superinfection.[14]

Chronic Hepatitis D and Its Natural Course

Chronic hepatitis D is considered a more severe form of chronic liver disease with rapid advancement to cirrhosis.[1,3,4] Recent epidemiological data from population based studies in HDV endemic areas have indicated a more benign course in a significant proportion of chronically infected individuals.[3]

Wu et al have proposed that patients with chronic HDV infection progress from an initial phase of active hepatitis to a chronic phase with moderate disease activity.[14] From that point, patients either progress to cirrhosis or remain in a chronic remission state characterized by decreased HDV and HBV replication.[14]

Development of cirrhosis has been reported in 70 to 80% of chronically infected patients (15% occurring during the first 2 years of infection).[15,16] The significant impact of HDV infection in the development of cirrhosis is best illustrated in HBsAg positive cirrhotic children where ~ 40% of them have evidence of HDV infection.[1] Following this rapid phase of cirrhosis development, it appears that the disease activity subsides with long periods of inactivity prior to the emergence of clinical signs of decompensated cirrhosis.[2] Nevertheless, during this period the risk of HCC development and death is increased by threefold and twofold, respectively compared to HDV negative cirrhotic patients with chronic HBV infection.[17] Once the stage of clinical cirrhosis has been reached, replication of HDV declines and HDV RNA is detected in approximately one third of patients.[2]

The clinical impact of chronic HDV infection in areas endemic for HDV infection is also apparent by the proportion of HBV infected individuals that undergo liver transplantation. In recent data from Italy, 36% of the HBV transplanted patients had a concomitant HDV infection.[18]

The ominous prognosis of chronic HDV/HBV infection has been recently reexamined in the light of recent epidemiological data showing a significant overall decrease in the prevalence of HDV infection in endemic areas such as Italy and Greece[3,19] as well as by the realization that this chronic infection does not always follow a progressively deteriorating course.[3]

In a recent multicenter Italian study the prevalence of HDV infection among patients with chronic HBV infection has declined from 23% in 1987 to 8% in 1997.[19] Similarly, among HBV cirrhotic patients the proportion of HDV infected individuals has dropped from 40% to 12% during the same period.[19] This decrease is attributed mainly to a significant decline in the number of new HBV/HDV infections while at the same time most cases of newly diagnosed chronic HDV infections presented at an advanced stage of liver disease (70%

cirrhotic).[2] Similar observations have been made in other endemic areas of chronic hepatitis D like Greece.[3]

On the other hand, studies in closed populations of highly endemic areas in Greece and the American Samoa, have shown that a significant proportion of patients with chronic HDV infection follow a benign course with minimal liver necroinflammatory activity and low rates of progression to cirrhosis.[3] Furthermore, increased rates of spontaneous clearance of HBsAg (10-20%) have been reported in patients with chronic HBV/HDV infection.[3,20] These clinical observations emphasize the significant heterogeneity in the clinical outcome of chronic HBV/HDV infection.

The exact determinants and adverse prognostic factors for progressive liver disease in patients with chronic HDV infection have not been clarified. Among the factors that are presumed to play a role is the HDV genotype, the level of HBV and HDV replication as well as the coinfection with other viruses (HCV, HIV). So far, three genotypes of HDV (I, II, III) have been reported. Although certain epidemiological observations regarding the natural course and frequency of fulminant hepatitis in the three different genotypes have been made, their exact role in the clinical outcome of chronic HDV infection has not been determined. Genotype I is the most common genotype worldwide including United States, Middle East and Europe. Recently, cases of fulminant hepatitis in patients infected with HDV genotype I have been identified in Russia (city of Samara).[13] Genotype II is more common in Asia (Japan, Taiwan) and is associated with milder forms of the disease. Genotype III has been isolated only in South America (Columbia, Peru, Venezuela), usually in association with genotype F HBV infection. In these areas, clusters of fulminant hepatitis D have been described.[21]

Coinfection with HCV and HIV

Triple infections with HBV, HDV and HCV can be identified in several populations at risk for multiple parenterally transmitted viral infections as is the case of intravenous drug users (IVDUs), multiply transfused individuals and HIV infected persons. The same is true for a Taiwanese community of Tzukuan where in addition to HBV and HCV endemicity, HDV infection was also found to be endemic probably because of sharing of nondisposable needles and syringes for daily medical practices.[22]

As mentioned above, in patients coinfected with HBV, HDV exhibits an inhibitory role on the replicative efficacy of HBV during the chronic phase of the disease.[22-24] In triple infections, in most studies HDV is the predominant virus suppressing the replication of HBV and HCV.[23-25] This was not the case in studies from Far East which demonstrated a predominant role for HCV in triple infections.[22,26] Regardless, of the predominating virus in triple infection a more severe course of liver disease has been reported.[24,25]

Although the data on chronic HDV/HIV coinfection are limited, it does not appear that there is an adverse effect of HIV infection in the course of liver disease in this group of patients.[27,28]

References

1. Farci P. Delta hepatitis: An update. J Hepatol 2003; 39(Suppl 1):S212-S219.
2. Rosina F, Conoscitore P, Cuppone R et al. Changing pattern of chronic hepatitis D in Southern Europe. Gastroenterology 1999; 117(1):161-166.
3. Hadziyannis SJ. Hepatitis D. Clin Liver Dis 1999; 3(2):309-325.
4. Rizzetto M, Smedile A. Hepatitis D. In: Schiff E, Sorrel M, Maddrey W, eds. Sciff's Diseases of the Liver. Lippincott Williams and Wilkins, 2002:863-875.
5. Sagnelli E, Stroffolini T, Ascione A et al. Decrease in HDV endemicity in Italy. J Hepatol 1997; 26(1):20-24.

6. Hadziyannis SJ. Viral Hepatitis: Clinical features. In: Bacon BR, Di Bisceglie AM, eds. Liver diseases: Diagnosis and Management. New York: Churchill Livingstone, 2000:79-107.

7. Caredda F, Antinori S, Pastecchia C et al. Incidence of hepatitis delta virus infection in acute HBsAg-negative hepatitis. J Infect Dis 1989; 159(5):977-979.

8. Chulanov VP, Shipulin GA, Schaefer S et al. Kinetics of HBV DNA and HBsAg in acute hepatitis B patients with and without coinfection by other hepatitis viruses. J Med Virol 2003; 69(3):313-323.

9. Buti M, Esteban R, Jardi R et al. Clinical and serological outcome of acute delta infection. J Hepatol 1987; 5(1):59-64.

10. Hadziyannis SJ, Dourakis SP, Papaioannou C et al. Changing epidemiology and spreading modalities of hepatitis delta virus infection in Greece. Prog Clin Biol Res 1993; 382:259-266.

11. Tang JR, Cova L, Lamelin JP et al. Clinical relevance of the detection of hepatitis delta virus RNA in serum by RNA hybridization and polymerase chain reaction. J Hepatol 1994; 21(6):953-960.

12. Casey JL, Brown TL, Colan EJ et al. A genotype of hepatitis D virus that occurs in northern South America. Proc Natl Acad Sci USA 1993; 90(19):9016-9020.

13. Flodgren E, Bengtsson S, Knutsson M et al. Recent high incidence of fulminant hepatitis in Samara, Russia: Molecular analysis of prevailing hepatitis B and D virus strains. J Clin Microbiol 2000; 38(9):3311-3316.

14. Wu JC, Chen TZ, Huang YS et al. Natural history of hepatitis D viral superinfection: Significance of viremia detected by polymerase chain reaction. Gastroenterology 1995; 108(3):796-802.

15. Saracco G, Rosina F, Brunetto MR et al. Rapidly progressive HBsAg-positive hepatitis in Italy. The role of hepatitis delta virus infection. J Hepatol 1987; 5(3):274-281.

16. Rizzetto M, Verme G, Recchia S et al. Chronic hepatitis in carriers of hepatitis B surface antigen, with intrahepatic expression of the delta antigen. An active and progressive disease unresponsive to immunosuppressive treatment. Ann Intern Med 1983; 98(4):437-441.

17. Fattovich G, Giustina G, Christensen E et al. Influence of hepatitis delta virus infection on morbidity and mortality in compensated cirrhosis type B. The European Concerted Action on Viral Hepatitis (Eurohep). Gut 2000; 46(3):420-426.

18. Fagiuali S, Mirante VG, Pompili M et al. Liver transplantation: The Italian experience. Dig Liver Dis 2002; 34(9):640-648.

19. Gaeta GB, Stroffolini T, Chiaramonte M et al. Chronic hepatitis D: A vanishing Disease? An Italian multicenter study. Hepatology 2000; 32(4 Pt 1):824-827.

20. Niro GA, Gravinese E, Martini E et al. Clearance of hepatitis B surface antigen in chronic carriers of hepatitis delta antibodies. Liver 2001; 21(4):254-259.

21. Casey JL, Niro GA, Engle RE et al. Hepatitis B virus (HBV)/hepatitis D virus (HDV) coinfection in outbreaks of acute hepatitis in the Peruvian Amazon basin: The roles of HDV genotype III and HBV genotype F. J Infect Dis 1996; 174(5):920-926.

22. Lu SN, Chen TM, Lee CM et al. Molecular epidemiological and clinical aspects of hepatitis D virus in a unique triple hepatitis viruses (B, C, D) endemic community in Taiwan. J Med Virol 2003; 70(1):74-80.

23. Jardi R, Rodriguez F, Buti M et al. Role of hepatitis B, C, and D viruses in dual and triple infection: Influence of viral genotypes and hepatitis B precore and basal core promoter mutations on viral replicative interference. Hepatology 2001; 34(2):404-410.

24. Sagnelli E, Coppola N, Scolastico C et al. Virologic and clinical expressions of reciprocal inhibitory effect of hepatitis B, C, and delta viruses in patients with chronic hepatitis. Hepatology 2000; 32(5):1106-1110.

25. Mathurin P, Thibault V, Kadidja K et al. Replication status and histological features of patients with triple (B, C, D) and dual (B, C) hepatic infections. J Viral Hepat 2000; 7(1):15-22.

26. Liaw YF, Tsai SL, Sheen IS et al. Clinical and virological course of chronic hepatitis B virus infection with hepatitis C and D virus markers. Am J Gastroenterol 1998; 93(3):354-359.

27. Monno L, Angarano G, Santantonio T et al. Lack of HBV and HDV replicative activity in HBsAg-positive intravenous drug addicts with immune deficiency due to HIV. J Med Virol 1991; 34(3):199-205.

28. Pol S, Wesenfelder L, Dubois F et al. Influence of human immunodeficiency virus infection on hepatitis delta virus superinfection in chronic HBsAg carriers. J Viral Hepat 1994; 1(2):131-137.

CHAPTER 8

Diagnosis of Hepatitis D Virus Infection

Jaw-Ching Wu*

Introduction

Hepatitis D virus (HDV) is a small defective virus with a single stranded circular RNA of 1.7 kb in size.[1-3] Its hepatitis B surface antigen (HBsAg) envelope is provided by the helper hepatitis B virus (HBV) for successful package and transmission of HDV.[4,5] The antigenomic strand of HDV encodes a single protein, hepatitis delta antigen (HDAg) of two molecular weight forms. The large form HDAg with a 19-amino acid extension at the C-terminus plays a key role in the package of HDV and suppresses viral replication in a trans-dominant negative manner, while the small form plays an essential role in trans-activating the replication of HDV RNA.[2,3]

There are two kinds of HDV infection.[4] Coinfection results from acute simultaneous infection of both HBV and HDV. Few patients with coinfection progress to chronicity because of self-limited nature of acute HBV infection in adults. Superinfection indicates the occurrence of HDV infection in patients with underlying chronic hepatitis B.[4] The great majority of patients with HDV superinfection progress to chronicity.[6] The disease spectrum of HDV infection varies greatly from an important etiology of fulminant hepatitis and rapidly progressive hepatitis to a subclinical course.[6-13] Persistent replication of HDV associated with relapsing acute exacerbations and elevated ALT levels is a characteristic of chronic active hepatitis D.[6] HDV is classified into three genotypes.[14] The most common isolate, the genotype I, has been cloned from patients with fulminant or chronic active hepatitis in Italy, America, Taiwan, Nauru, France and Lebanon. Genotype II has only been isolated from Japan and Taiwan, and it is less often associated with fulminant hepatitis or rapid progression to cirrhosis or HCC as compared to genotype I.[15,16] Genotype III has been isolated from patients with severe acute hepatitis in Peru and Colombia.[14]

Several steps are needed for an accurate diagnosis of HDV infection. The first step is to differentiate HBV from HDV infection. Differential diagnosis is essential for understanding and proper management of the disease, because HDV infection is an important etiology of fulminant hepatitis and an aggravating factor for the progression of chronic hepatitis B to cirrhosis.[6-12] In the clinical setting of antiviral therapy, nucleoside analogues are effective in the treatment of chronic hepatitis B, but not effective for chronic hepatitis D. In the setting of liver transplantation, recurrent HBV infection is associated with inflammation and tissue damage of the graft and the mortality of recipients, whereas HDV infection in the host liver seems to

*Jaw-Ching Wu—Division of Gastroenterology, Taipei Veterans General Hospital, Insitute of Clinical Medicine, National Yang-ming University, 201 Shih-Pai Road, Sec. 2, Taipei 112, Taiwan. Email: jcwu@vghtpe.gov.tw

Hepatitis Delta Virus, edited by Hiroshi Handa and Yuki Yamaguchi.
©2006 Landes Bioscience and Springer Science+Business Media.

Table 1. Serological and histological markers for the diagnosis of HDV infection

Markers	Methods	Comments
Serum		
Total anti-HDV	Immunoassay	Positive in coinfection and superinfection
		High titer in active infection
IgM anti-HDV	Immunoassay	Positive in acute and chronic infection
		High titer in active infection
HDV RNA	Northern blot	Active HDV replication
	RT-PCR	Most sensitive, 10-100 copies
HDAg	Immunoassay	Masked in immune complex in chronic infection
	Western blot	Active HDV replication, research use
HBsAg	Immunoassay	Usually positive
IgM anti-HBc	Immunoassay	Positive in coinfection
Liver		
HDV RNA	Northern blot	Active replication, research use
	In situ hybridization	Active replication, research use
HDAg	Western blot	Active replication, research use
	Immuno-staining	Active replication, classical standard

reduce the risk of significant HBV-induced liver damage in the allograft.[17] The survival rate of liver transplant recipients is better in patients with both HBV and HDV infections than that of patients with HBV infection alone.[17] The different outcomes may be due to suppressed HBV replication in patients with chronic hepatitis D. The second step is to distinguish between HDV coinfection and superinfection, because the former seldom progresses to chronicity and the great majority of the latter becomes chronic. The third step is the differentiation between acute and chronic hepatitis D. Finally, the determination of HDV genotypes is also of clinical and epidemiological importance.[14,16] The diagnosis of HDV infection is based on the detection of the components (HDAg or HDV RNA) of HDV or antibodies to HDAg that will be discussed in the following sections (Table 1).

Serological Diagnosis Based on Antibodies to HDAg (Anti-HDV)

Radio- or enzyme-immunoassays of serum anti-HDV are commercially available, and are most convenient for the first-line screening of HDV infection in a large number of patients in daily clinical practice or epidemiological surveys. Acute HDV infection is diagnosed by seroconversion or rising titers of anti-HDV.[18,19] At acute stage, some patients with acute HDV coinfection or superinfection are diagnosed by seroconversion of anti-HDV.[19,20] Therefore, follow-up assays of anti-HDV at one month interval are needed to determine if HDV infection is the etiology responsible for acute hepatitis attacks in patients, particularly in those with risk behaviors (intravenous drug abuse or prostitute contact). However, patients with fulminant hepatitis may expire before seroconversion of anti-HDV. In such cases, detection of HDV RNA by a sensitive reverse transcription polymerase chain reaction (RT-PCR) is of great value to determine the etiology of fulminant hepatitis.[9] Within 2 months of HDV infection, more than 90% of patients become serum anti-HDV positive. Of the patients whose initial serum samples are already positive for anti-HDV, the diagnosis of acute HDV infection may be supported by rising titers of anti-HDV. If only one serum sample is tested, titration of anti-HDV is of value in the differential diagnosis between acute and chronic hepatitis D. The highest dilution that gives positive result is defined as anti-HDV titer. A low anti-HDV titer less than

Table 2. Differential diagnosis of HDV infection

Serum	Acute		Chronic	
	Coinfection	Superinfection	Active	Remission
Serum				
HBsAg	Transient	Persistent	Persistent	Persistent
IgM anti-HBc	Positive	Negative	Negative	Negative
IgG anti-HDV*	Transient	Initial <100	Often >1000	Often<1000
IgM anti-HDV	Transient	Often positive	Often positive	Often negative
HDV RNA	Transient	Often positive	Often positive	Often negative
ALT	High[†]	High[†]	Moderate	Normal
Liver				
HDAg	Transient	Often positive	Often positive	Often negative
HDV RNA	Transient	Often positive	Often positive	Often negative

* Total anti-HDV titer: Highest dilution of serum that still shows positive anti-HDV testing. [†] Usually > 10 X normal.

100 in patients with acute hepatitis attack is a characteristic finding of acute HDV superinfection.[19,21] The sensitivity is 98% and the specificity is 93%.[19]

The differentiation between coinfection and superinfection is important. Less than 5% of HDV coinfections progress to chronicity due to self-limited nature of acute hepatitis B in adults. On the contrary, more than 90% of HDV superinfections become chronic.[6,20,21] Both coinfection and superinfection of HDV may induce fulminant hepatic failure.[6,20,21] The mortality rate of HDV coinfection ranges from 1-10%, while that of HDV superinfection may be up to 5-20%.[21] Coinfection is characterized by coexistence of high titer IgM anti-HBc and IgM anti-HDV (Table 2). Anti-HDV appears at 1 week after infection, and IgM anti-HDV usually disappears within 6 weeks. IgM anti-HDV may occasionally persist up to 12 weeks.[18] Persistence of IgM anti-HDV predicts the progression to chronic HDV infection, while clearance of IgM anti-HDV often indicates resolution of infection. In the cases of HDV superinfection, serum IgM anti-HBc is usually negative or occasionally in low titer. In contrast to IgM anti-HBc for acute hepatitis B, high titer IgM anti-HDV often persists in chronic hepatitis D and is associated with active disease activity. Jardi et al reported that high molecular weight IgM form (19 S) was predominantly detected in patients with acute HDV infection, whereas the low molecular weight (7 S) form was found in chronic hepatitis D cases.[22] However, there are some overlaps of the results. And the procedures for the separation and the determination of the 17-S and 19-S forms of IgM anti-HDV are not convenient for routine clinical use.

The role of serum anti-HDV in representing active HDV infection and liver disease activity is controversial.[23-28] Early studies defined a high serum titer (≥1000 or ≥5000) of total anti-HDV to represent chronic hepatitis D (CHD).[23-25] Buti et al reported that IgM anti-HDV was associated with active HDV replication and disease activity.[23] In contrast, a lack of correlation of IgM anti-HDV with HDV replication and histological activity was reported in later studies.[19,26-28] In our study,[19] a high serum titer (≥1000 or ≥5000) of total anti-HDV has a high positive predictive rate (84-86%) for HDV viremia, but the negative predictive rate (36-46%) is low (Table 3). Serum IgM anti-HDV also has a high positive predictive rate of 89% for HDV viremia, but the negative predictive rate (54%) is also low (Table 2). In another study using hepatic HDAg staining as a reference of HDV replication by Jardi et al[28] serum HDV RNA had the highest concordance rate, followed by serum HDAg, and serum IgM

Table 3. Correlation of serological and histological markers with HDV viremia based on RT-PCR in chronic anti-HDV positive patients

Indices	Sensitivity	Specificity	PP/ NP	κ value
Serum anti-HDV titer				
≥ 5000	0.15	0.95	0.86 / 0.36	0.07
≥ 1000	0.52	0.80	0.84 / 0.46	0.27
≥ 100	0.94	0.43	0.76 / 0.77	0.41
Serum IgM Anti-HDV	0.62	0.85	0.89 / 0.54	0.41
HDV-RNA by northern blot	0.71	1.00	1.00 / 0.78	0.72
Hepatic HDAg staining	0.82	0.94	0.97 / 0.71	0.70

PP: Positive predictive rate, NP: Negative predictive rate. (Reprinted with permission from ref. 19, Huang et al, J Gastroenterol Hepatol 1998; 13:57-61. ©1998 Blackwell Publishing.)

determination had the least degree of concordance. Discrepant results may be also observed in immuno-compromised patients who have HDV viremia in the absence of serum anti-HDV. On the other hand, some patients with HDV coinfection or superinfection may still have detectable anti-HDV in sera after clearance of serum HDV or even HBsAg. The anti-HDV titers of these patients are usually low.

Molecular Diagnosis Based on HDAg and HDV RNA

Detection of HDAg or HDV RNA in serum or liver indicates active HDV replication and infectivity.[6,12,26,28-37] The following sections will describe the applications and limitations of various methods in the detection of HDAg or HDV RNA.

Serum HDAg

Detection of HDAg in serum by enzyme immunoassay or western blot represents active HDV replication.[28] Serum HDAg and anti-HDV IgM are sensitive for diagnosis of acute HDV infection during the first two weeks after onset of symptoms.[20] However, HDAg is undetectable by immunoassay 2 weeks after infection, probably because the anti-HDV in serum has formed immunocomplex with HDAg and competes for anti-HDV in the immunoassays. The detection rates of HDAg by immunoassays in patients with acute HDV infection vary from 26% to 100%.[36,37] Timing of blood sampling and different sensitivity in the assays may explain the discrepant results. The detection of serum HDAg by enzyme immunoassay has not been widely applied in clinical practice.

In contrast to enzyme immunoassay, immunoblotting is sensitive in the detection of HD Ag either in the sera of patients with acute or chronic HDV infection. Before Western blot analysis of HDAg, serum samples are firstly ultracentrifuged and viral particles can be pelleted through sucrose cushion. Then viral proteins are extracted and subjected to polyacrylamide gel electrophoresis and transferred to nitrocellulose membrane. The blotted HDAg can react with primary antibody of anti-HDV, followed by secondary antibody labeled with radioactive isotope, immuno-peroxidase or chemiluminescence agent. Finally, HDAg will be shown by autoradiography, chloromogens or chemiluminescence. The HDAgs in viral particles are shown as the small form of 24 kDA and the large form of 27 kDA, approximately in equal amounts, in contrast to larger amount of small HDAg in liver tissue. Using this method, serum HDAg was detected in 71% of chronic hepatitis D patients with intrahepatic HDAg.[28] However, this method is cumbersome for routine clinical purpose and less sensitive compared to RT-PCR or

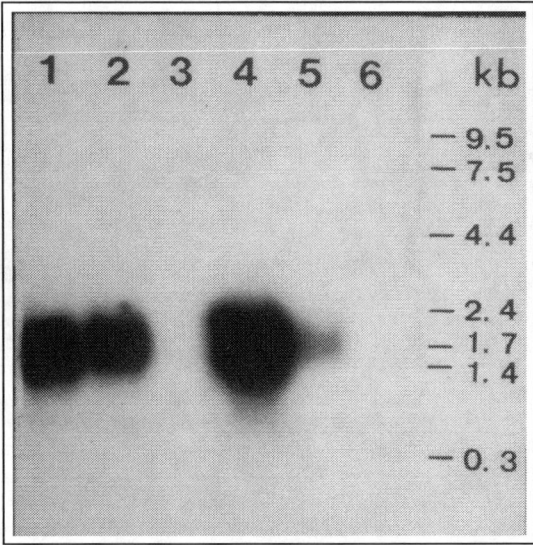

Figure 1. Northern blot hybridization using strand-specific riboprobe for full-genomic sequence. Lanes 1-4 are anti-HDV positive patients with acute hepatitis. Lane 5 is a patient with chronic hepatitis D. Lanes 1, 2, and 4 show strong positive signals at 1.7 kb of HDV genomic sequence. Lane 5 shows a weaker signal at 1.7 kb. Lane 3 shows negative result. Lane 6 is a negative control.

Northern blot hybridization using riboprobe.[31] Smedile et al reported that the detection rate of serum HDV RNA using riboprobe or cDNA probe, or serum HDAg by immunoblotting in chronic hepatitis D patients with intrahepatic HDAg were 83%, 63% and 73%, respectively.[31]

Serum HDV RNA

Serum HDV RNA can be detected either by dot-blot hybridization, Northern blot hybridization or RT-PCR. Viral particles are ultracentrifuged through sucrose cushion to form pellets at the bottom of tubes. Then viral RNA is extracted and analyzed in agarose gel electrophoresis. Extracted HDV RNA can be directly spotted or transferred after electrophoresis onto nitrocellulose membrane, and hybridized with a cDNA or riboprobes. The Northern blot hybridization not only shows positive signal but also demonstrate the correct molecular weight of HDV genome at 1.7 kb (Fig. 1). Therefore, Northern blot hybridization provides a specific signal that can be more easily read. Incomplete denaturation of HDV RNA before Northern blot hybridization may yield additional bands besides the correct signal at 1.7 kb. Well-prepared and -preserved HDV RNA from the degradation of ribonuclease, freshly-prepared acid phenol for RNA extraction and a riboprobe with high specific activity are key points for the assurance of high sensitivity of Northern blot hybridization. The sensitivity of Northern blot detection of HDV RNA using RNA probe is superior to that using cDNA probe. The use of full-length riboprobes may further increase the sensitivity compared to the use of cDNA or RNA probes of partial genomic sequences.[6,19,31] Moreover, strand-specific riboprobes can detect either genomic and anti-genomic HDV RNA in infected or transfected hepatocytes and thus confirm the replication of HDV.[5] Because HDV genome has high GC contents and intra-molecular complementarity,[1-3] stringent washing conditions and temperatures are required to assure a clean background. There are no commercially available kits for the quantification of HDV RNA. Comparison of the signals of slot-hybridization or Northern blot hybridization with

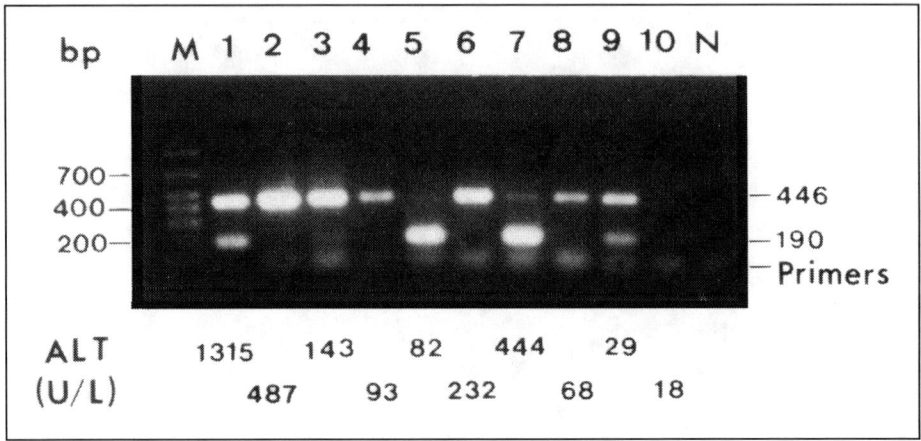

Figure 2. Detection of serum HBV DNA by PCR and HDV RNA by RT-PCR in patients with HDV infection at various stages of infection. The PCR or RT-PCR products of the same samples were loaded into the same wells, electrophoresed in a 2% agarose gel, and stained with ethidium bromide. The 190-base pair bands represented amplified cDNA products of the core gene of HBV, and the 446-base pair bands represented the amplified cDNA products of the HDAg-coding region of HDV. The lowest bands represented the remnant of the primers used in PCR. M: indicates molecular size markers shown by a base pair (bp). Lanes 1-8 represent the amplified HDV products of acute nonfulminant hepatitis, chronic hepatitis, cirrhosis, and HCC with or without accompanying HBV products. Lane 9 indicates a remission case with positive HBV DNA and HDV RNA, and lane 10 indicates the other remission case negative for both. N indicates negative samples from HBsAg-negative and anti-HDV-negative healthy control. (Reprinted with permission from Wu et al, Gastroenterology 1995; 108:796-802, ©1995 American Gastroenterological Association.)[6]

hybridized signals of serial dilutions of in vitro transcribed HDV RNA or cDNA of HDV of know concentrations can provide relative amounts of serum HDV RNA of patients. Cariani et al developed a nonradioisotopic assay for the detection of HDV RNA in serum by combining RT-PCR of the RNA and enzyme linked immunoassay detection of the PCR products using a monoclonal antibody specific for double-stranded DNA.[38] The sensitivity of this method was comparable to that of standard PCR followed by Southern-blot hybridization (about 10 copies/ml) and was 1000 to 10000 times more sensitive than direct dot-blot hybridization.[38] The determination of serum HDV RNA levels of patients can be used to study the changes of HDV replication during the natural history of HDV infection.[6] The replication of HDV is more active at the acute stage of infection, then gradually decreasing afterwards.[6] The persistence of HDV replication is usually associated with elevated ALT levels, and the clearance of HDV RNA is often accompanied by normalization of ALT levels unless HBV reactivates.[6] However, some asymptomatic risk groups may have HDV viremia in the absence of elevated ALT levels.[39] The measurement of serum HDV RNA levels has been also used to evaluate the therapeutic response of patients with chronic hepatitis D.[38,40]

RT-PCR (Fig. 2) has been documented as the most sensitive assay for HDV-RNA detection (Table 3). It is 10,000 times more sensitive than northern blot hybridization.[6,28-34,38] Micro-heterogeneity in nucleotide sequences of different HDV clones from a single subject has been reported.[41] HDV is classified into three genotypes including a novel subtype of genotype II.[14-16,41,42] The divergence in nucleotide sequences ranges from 5% to 14% among different isolates of the same genotype and from 22% to 38% among different genotypes.[14,15,41,42] It has

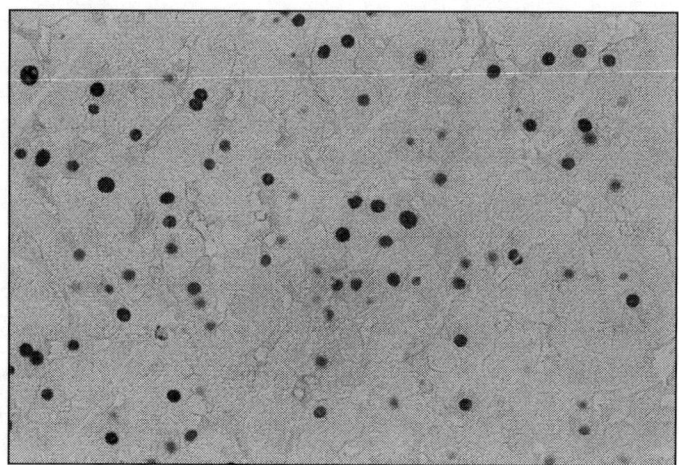

Figure 3. Immuno-peroxidase staining of HDAg in liver biopsy of a patient with acute HDV superinfection. Diffuse distribution of nuclear staining (predominantly both nucleoplasmic and nucleolar staining and some nucleoplasmic staining only) in more than 50% of hepatocytes (original X 100).

been suggested that faster rate of evolution in the RNA viral genome may be due to the low proof-reading activity of RNA polymerase.[15,43] Recombination may further increase the diversity of HDV genomes.[44] The heterogeneity of HDV genomic sequences among different isolates and genotypes and the high intramolecular complementarity with secondary or tertiary structures of HDV genomes may reduce the efficiency of RT-PCR unless experiments are carried out in optimal conditions.[1-3,6,14,15,42-44] Proper extraction and preservation of HDV RNA genomes to be used as templates in reverse transcription is the first step to assure a high sensitivity. Second, primers for PCR should be synthesized based on conserved regions of the prevalent HDV genotypes of that area.[6,14,15,42-44] Primers based on the HDAg-coding region, particularly the sequence encoding the carboxyl half of the HDAg, can give a high yield of PCR.[6,14,42-44] Using RT-PCR based on these primers, HDV RNA can be detected in the sera of 97% patients with acute HDV superinfection, 75% of chronic hepatitis patients with positive anti-HDV, 74% of anti-HDV positive cirrhotic patients, 63% of anti-HDV positive patients with hepatocellular carcinoma, and 34% of anti-HDV positive subjects with biochemical remission.[6] Of the patients negative for serum HDV RNA by dot-blot or Northern- blot hybridization, a significant portion (>54%) still had detectable HDV RNA[6,19,33] Quantitative PCR can be used for the determination of HDV RNA levels in the sera of patients.

Immuno-Pathologic Diagnosis Based on Hepatic HDAg

The demonstration of nuclear staining of a novel antigen in liver by immunofluoresence technique led to the first discovery of HDAg by Dr. Rizetto et al[45] Positive detection of HDAg in liver biopsies by immuno-peroxidase (Fig. 3) or immuno-fluorescence staining still remain to be important diagnostic methods for HDV infection.[45-48] The immuno-peroxidase staining can be applied to paraffin-embedded liver specimens without the requirement of fresh-frozen tissues, and therefore is more convenient for routine use. Moreover, immuno-peroxidase staining in paraffin-embedded liver specimens provides a better general outline of hepatocytes and inflammatory cells for histological inspection. Patient's sera with high titers of anti-HDV are common sources of primary antibody for HDAg staining. However, antibodies obtained from animals immunized with recombinant HDAg become more stable sources and provide

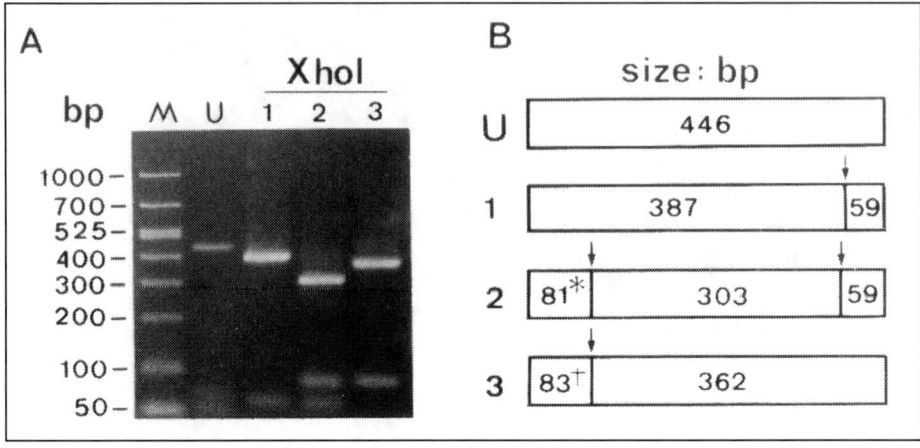

Figure 4. HDV genotyping by RFLP and the discovery of a novel HDV genotype. A) The *Xho*I-cleaved PCR products of HDV genomes were electrophoresed in a 3% agarose gel and stained with ethidium bromide. Lanes: M: molecular size markers (bp); U: undigested amplified products of 446 bp; 1: digested genotype I PCR products; 2: digested genotype II PCR; and 3, a novel RFLP pattern. B) restriction sites of *Xho*I in the amplified sequence (nt 889-1334) of different HDV genotypes. Vertical arrows indicate *Xho*I cutting sites. Horizontal bar: U: undigested amplified sequence which varied from 443 to 446 bp in size among different genotypes (genotype III HDV does not have any *Xho*I cutting site in this region); 1: genotype I PCR products have a single *Xho*I cutting site and are cleaved into fragments of 387 and 59 bp; 2: genotype II PCR products have two *Xho*I cutting sites and are cleaved into fragments of 81, 303 and 59 bp (*, the genotype II PCR products have three single-base deletions at 5' end of the amplified sequence); 3: PCR products of the novel genotype HDV isolate have a single *Xho*I cutting site and are cleaved into fragments of 83 and 362 bp (†, the novel HDV isolate has a single-base deletion at 5' end). (Ref. 42, Wu et al, J General Virol 1998; 79:1105-1113.)

constant titers with lower background staining. The demonstration of HDAg in liver biopsies indicates active HDV replication and has an excellent correlation with serum HDV RNA (Table 3).[6,19]

The HDAg is usually shown in the nuclei of hepatocytes of patients. With higher power field inspection, whole nuclei or nucleoplasm are the main subcellular localizations of HDAg staining (Fig. 4). In transfected culture cells,[49-52] nucleolar, nucleoplasmic or cytoplasmic staining can be found at different times after transfection. However, nucleolar or cytoplasmic staining patterns are rarely found in liver biopsy specimens of patients.[45-48,53] The finding of these two patterns at the very early stage of transfection may explain their absence in liver tissues of HDV-infected humans or animals that are obtained long after HDV infection.[45-53] The exact cause remains to be further clarified. The numbers of HDAg-staining nuclei vary from more than 50% of hepatocytes to a few hepatocytes in whole liver section.[6,48,53] During the natural history of HDV infection, there is a trend of decreasing percentage of HDAg-staining hepatocytes from acute to later stages, and from active liver disease to remission. The reduction of HDAg-staining hepatocytes generally parallels the decreasing serum HDV RNA during natural history.[6]

Genotypic Diagnosis

Because HDV genotypes appear to correlate with viral assembly efficiencies and clinical outcomes, determination of HDV genotypes is therefore of clinical importance.[14,16,54] The gold standard of genotyping of HDV is phylogenetic analysis based on sequence comparison among different HDV isolates.[14,42,44] A simpler method for rapid genotypic diagnosis in a large number of samples has been developed by analyzing the restriction fragment length polymorphism (RFLP) of RT-PCR products cleaved by a restriction enzyme *Xho*I (Fig. 4).[16] This method is also valuable in the discovery of novel HDV genotypes and the diagnosis of mixed genotypes infection of HDV.[42,55] Genotyping of HDV can be also established based on immunohistochemical differentiation using genotype-specific antibodies against the fusion proteins containing the carboxyl-terminal 19 amino acids of the large HDAg that have the highest divergent sequence among different genotypes.[14,41,42,56] These antibodies are also useful to differentiate the large HDAg from the small form because the latter does not have the carboxyl-terminal 19 amino acids of the large HDAg. However, this method is still limited in research use because genotype-specific antibodies are not commercially available at present.

Replication Markers of HBV

Because the assembly and transmission of HDV requires the provision of hepatitis B surface antigen envelope from the helper HBV, both coinfection and superinfection of HDV require the presence of HBV.[4,5] Furthermore, the status of HBV replication may modulate disease course and contribute to inflammatory activity of liver.[6,57] Relapse of liver disease activity in transplanted liver usually occurs after the reactivation of HBV. Lamivudine is highly effective for the treatment of chronic hepatitis B with active HBV replication,[58] but not for chronic hepatitis D patients with HDV viremia with or without coexistent HBV replication.[59] Theoretically, lamivudine may be effective for anti-HDV positive patients who have HBV replication in the absence of active replication of HDV. Therefore, testing of serological and replication markers of HBV is essential for correct diagnosis and proper management. There are several commercially available kits for the measurement of serum HBV DNA levels. HBV is usually suppressed at the acute stage of HDV infection, HBsAg may be transiently undetectable in the sera of some patients who are diagnosed as nonB, nonC hepatitis unless follow-up assays of HBsAg and anti-HDV are tested. Serum HBV DNA is usually negative or very low in concentrations at the acute stage of HDV infection.[6] It may reactivate in chronic HDV infection, but is mostly in low level. Nevertheless, HBV may be the sole marker of viral replication in the absence of HDV RNA, and account for inflammatory activity of the liver.[6] Sustained remission is usually found in patients with sero-clearance of both HBV and HDV.[6]

Summary

The possibility of concordant HDV infection should be taken into consideration in patients with acute or chronic hepatitis B, particularly in patients with intravenous drug abuse or promiscuity or sexual contacts of a known HDV-infected subject.[41] The diagnosis of HDV infection includes serological markers of both HBV and HDV to differentiate coinfection from superinfection. Testing of total anti-HDV is the most convenient and commercially available first-line assay to screen HDV infection. IgM anti-HBc in addition to anti-HDV is a diagnostic marker of HBV and HDV coinfection. IgM anti-HDV has some indirect correlation with HDV replication and inflammatory activity of the liver, but discrepant results do exist. Detection of HDAg and HDV RNA in serum or liver tissue provides evidences of active HDV replication that usually links to persistent hepatic inflammation. These assays are valuable for understanding the clinical course, predicting outcomes and evaluating antiviral treatment responses. Of these assays, RT-PCR is the most sensitive. Finally, genotyping of HDV is of clinical and molecular-epidemiological importance.

References

1. Wang KS, Choo KL, Weiner AJ et al. Structure, sequence and expression of the hepatitis delta (£δ) viral genome. Science 1986; 323:508-512.
2. Modahl LE, Lai MMC. Hepatitis delta virus: The molecular basis of laboratory diagnosis. Critical Rev Clin Lab Sciences 2000; 37:45-92.
3. Taylor JM. Replication of human hepatitis delta virus: Recent developments. Trends Microbiol 2003; 11:185-190.
4. Rizzetto M, Canese MG, Gerin JL et al. Transmission of hepatitis B virus-associated delta antigen to chimpanzee. J Infect Dis 1980; 141:590-602.
5. Wu JC, Chen PJ, Kuo MYP et al. Production of hepatitis D virus and suppression of helper hepatitis B virus in a human hepatoma cell line. J Virol 1991; 65:1099-1104.
6. Wu JC, Chen TZ, Huang YS et al. Natural history of hepatitis D viral superinfection - significance of viremia detected by polymerase chain reaction. Gastroenterology 1995; 108:796-802.
7. Hadler SC, De Monzo M, Ponzetto A et al. Delta virus infection and severe hepatitis: An epidemic in Yucpa Indians of Venezuela. Ann Intern Med 1984; 100:339-44.
8. Govindarajan S, Chin KP, Redeker AG et al. Fulminant B viral hepatitis: Role of delta antigen. Gastroenterology 1984; 86:1417-1420.
9. Wu JC, Chen CH, Hou MC et al. Multiple viral infections as the most common cause of fulminant and subfulminant viral hepatitis in an endemic area for hepatitis B: Application and limitations of polymerase chain reaction. Hepatology 1994; 19:836-840.
10. Rizzetto M, Verme G, Recchia S et al. Chronic hepatitis in carriers of hepatitis B surface antigen, with intrahepatic expression of the delta antigen. An active and progressive disease unresponsive to immunosuppressive treatment. Ann Intern Med 1983; 98:437-441.
11. Govindarajan S, De Cock KM, Redeker AG. Natural course of delta superinfection in chronic hepatitis B virus-infected patients: Histologic study with multiple liver biopsies. Gastroenterology 1986; 6:640-644.
12. Wu JC, Lee SD, Govindarajan S et al. Correlation of serum delta RNA with clinical course of acute delta virus superinfection in Taiwan: A longitudinal study. J Infect Dis 1990; 161:1116-1120.
13. Hadziyannis SJ, Hatzakis A, Paipaioannou C et al. Endemic hepatitis delta virus infection in a Greek community. Prog Clin Biol Res 1987; 234:181-202.
14. Casey JL, Brown TL, Colan EJ et al. A genotype of hepatitis D virus that occurs in northern South America. Proc Acad Sci USA 1993; 90:9016-9020.
15. Imazeki F, Omata M, Ohto M. Heterogeneity and evolution rates of delta virus RNA sequences. J Virol 1990; 64:5594-9.
16. Wu JC, Choo KB, Chen CM et al. Genotyping of hepatitis D virus by restriction fragment length polymorphism and its correlation with outcomes of hepatitis D. Lancet 1995; 346:939-941.
17. Lucey MR, Graham DM, Martin P et al. Recurrence of hepatitis B and delta hepatitis after orthotopic liver transplantation. Gut 1992; 33:1390-1396.
18. Argona M, Macagno S, Carreda F et al. Serological response to the hepatitis delta virus in hepatitis D. Lancet 1987; 1:478-480.
19. Huang YH, Wu JC, Sheng WY et al. Diagnostic value of anti-hepatitis D virus (HDV) antibodies revisited: A study of total and IgM anti-HDV compared with detection of HDV-RNA by polymerase chain reaction. J Gastroenterol Hepatol 1998; 13:57-61.
20. Buti M, Esteban R, Jardi R et al. Clinical and serological outcome of acute delta infection. J Hepatol 1987; 5:50-64.
21. Hoofnagle JH. Type D (Delta) hepatitis. JAMA 1989; 261:1321-1325.
22. Jardi R, Buti M, Rodriguez-Frias F et al. Clinical significance of two forms of IgM antibody to hepatitis delta virus. Hepatology 1991; 14:25-28.
23. Buti M, Esteban R, Esteban JI et al. Anti-HD IgM as a marker of chronic delta infection. J Hepatol 1987; 4:62-5.
24. Govindarajan S, Smedile A, Cock KMD et al. Study of reactivation of chronic hepatitis delta infection. J Hepatol 1989; 9:204-8.
25. Farci. P, Gerin JL, Aragona M et al. Diagnostic and prognostic significance of the IgM antibody to the hepatitis delta virus. JAMA 1986; 255:1443-6.

26. Govindarajan S, Gupta S, Valinluck B et al. Correlation of IgM anti-hepatitis D virus (HDV) to HDV RNA in sera of chronic HDV. Hepatology 1989; 10:34-5.
27. Lau JYN, Smith HM, Chaggar K et al. Significance of IgM anti hepatitis D virus (HDV) in chronic HDV infection. J Med Virol 1991; 33:273-6.
28. Jardi R, Buti M, Rodriguez F et al. Comparative analysis of serological markers of chronic delta infection: HDV RNA, serum HDAg and anti-HD IgM. J Virol Methods 1994; 50:59-66.
29. Smedile A, Rizzetto M, Denniston K et al. Type D hepatitis: The clinical significance of hepatitis D virus RNA in serum as detected by a hybridization-based assay. Hepatology 1986; 6:1297-1302.
30. Buti M, Esteban R, Roggendorf M et al. Hepatitis D virus RNA in acute delta infection: Serological profile and correlation with other markers of hepatitis D virus infection. Hepatology 1988; 8:1125-1129.
31. Smedile A, Bergmann KF, Baroudy BM et al. Riboprobe assay for HDV RNA: A sensitive method for the detection of the HDV genome in clinical serum sample. J Med Virol 1990; 30:20-24.
32. Tang JR, Cova L, Lamelin JP et al. Clinical relevance of the detection of hepatitis delta virus RNA in serum by RNA hybridization and polymerase chain reaction. J Hepatol 1994; 21:953-60.
33. Madejon A, Castillo I, Bartolome J et al. Detection of HDV-RNA by PCR in serum of patients with chronic HDV infection. J Hepatol 1990; 11:381-4.
34. Jardi R. Buti M, Cotrina M et al. Determination of hepatitis delta virus RNA by polymerase chain reaction in acute and chronic delta infection. Hepatology 1995; 21:25-9.
35. Negro F, Bonino F, Bisceglie AD et al. Intrahepatic markers of hepatitis delta virus infection: Study by in situ hybridization. Hepatology 1989; 10:916-920.
36. Govindarajan S, Valinluck B, Peters RL. Relapse of acute B viral hepatitis: Role of delta agent. Gut 1996; 27:19-22.
37. Shattock AG, Morgan BM. Sensitive enzyme immunoassasy for the detection of delta antigen and anti-delta, using serum as the delta antigen source. J Med Virol 1984; 13:73-82.
38. Cariani E, Ravaggi A, Puoti M et al. Evaluation of hepatitis delta virus RNA levels during interferon therapy by analysis of polymerase chain reaction products with a nonradioisotopic hybridization assay. Hepatology 1992; 15:685-689.
39. Wu JC, Li CS, Chen CL et al. Factors associated with viremia and elevated transaminase levels in asymptomatic hepatitis D virus-infected risk groups. J Med Virol 1994; 42:86-90.
40. Farci P, Mandas A, Colana A et al. Treatment of chronic hepatitis D with interferon alfa-2a. N Engl J Med 1994; 330:88-94.
41. Wu JC, Chen CM, Sheen IJ et al. Evidence of transmission of hepatitis D virus to spouses from sequence analysis of the viral genome. Hepatology 1995; 22:1656-1660.
42. Wu JC, Chiang TY, Sheen IJ. Characterization and Phylogenetic analysis of a novel hepatitis D virus group discovered by restriction fragment length polymorphism. J General Virol 1998; 78:1105-1113.
43. Lee CM, Bih FY, Chao YC et al. Evolution of hepatitis delta virus RNA during chronic infection. Virology 1992; 188:265-73.
44. Wu JC, Chiang TY, Shiue WK et al. Recombination of hepatitis D virus and its implications. Mol Biol Evol 1999; 16:1622-1632.
45. Rizzetto M, Canese MG, Arico S et al. Immunofluorescence detection of a new antigen/antibody system (delta-antidelta) associated with the hepatitis B virus in the liver and in the serum of HBsAg carriers. Gut 1977; 18:997-1003.
46. Recchia S, Rizzii R, Acquaviva F. Immunoperoxidase staining of the HBV-associated delta antigen in paraffinated liver specimens. Pathologica 1981; 73:773-777.
47. Stocklin E, Gudat F, Krey G et al. δ-antigen in hepatitis B: Immunohistology of frozen and paraffin-embedded liver biopsies and relation to HBV infection. Hepatology 1981; 1:238-242.
48. Kanel GC, Govindarajan S, Peters RL. Chronic delta infection and liver biopsy changes in chronic active hepatitis B. Ann Intern Med 1984; 101:51-54.
49. Kuo MYP, Chao M, Taylor J. Initiation of replication of the human hepatitis delta virus genome from cloned DNA: Role of delta antigen. J Virol 1989; 63:1945-1950.
50. Macnaughton TB, Gowans EJ, Jiber AR et al. Hepatitis delta virus RNA, protein synthesis and associated cytotoxicity in a stably transfected cell line. Virology 1990; 177:696-698.

51. Wu JC, Chen CL, Lee SD et al. Expression and localization of the small and large delta antigens during the replication cycle of hepatitis D virus. Hepatology 1992; 16:1120-1127.
52. Bichko VV, Taylor JM. Redistribution of the delta antigens in cells replicating the genome of hepatitis delta virus. J Virol 1996; 70:8064-8070.
53. Hadziyannis SJ, Sherman M, Lieberman HM et al. Liver disease activity and hepatitis B virus replication in chronic delta antigen-positive chronic hepatitis B virus carriers. Hepatology 1985; 5:544-547.
54. Hsu SC, Syu WJ, Sheen IJ et al. Varied Assembly and RNA Editing Efficiencies between Genotypes I and II Hepatitis D Virus and their Implications. Hepatology 2002; 35:665-672.
55. Wu JC, Huang IA, Huang YS et al. Mixed genotypes infection of hepatitis D virus. J Med Virol 1999; 57:64-67.
56. Hsu SC, Syu WJ, Ting LT et al. Immunohistochemical differentiation of hepatitis D virus genotypes. Hepatology 2000; 32:1111-1116.
57. Smedile A, Rosina F, Saracco G et al. Hepatitis B virus replication modulates pathogenesis of hepatitis D virus in chronic hepatitis D. Hepatology 1991; 13:413-416.
58. Lai CL, Chien RN, Leung NW et al. A one-year trial of lamivudine for chronic hepatitis B. Asian hepatitis lamivudine study group. N Engl J Med 1998; 339:61-68.
59. Lau DT, Doo E, Par K et al. Lamivudine for chronic delta hepatitis. Hepatology 1999; 30:579-581.

Index